T0270911

Topics in the Constructive Theory of

Countable Markov Chains

Topics in the Constructive Theory of Countable Markov Chains

G. Fayolle

INRIA

V.A. Malyshev

INRIA

M.V. Menshikov

Moscow State University

Published by the Press Syndicate of the University of Cambridge
The Pitt Building, Trumpington Street, Cambridge CB2 1RP
40 West 20th Street, New York, NY 10011-4211, USA
10 Stamford Road, Oakleigh, Melbourne, 3166, Australia

www.cambridge.org
Information on this title: www.cambridge.org/9780521461979

First published 1995

Library of Congress cataloguing in publication data available
A catalogue record for this book is available from the British Library

ISBN-13 978-0-521-46197-9 hardback
ISBN-10 0-521-46197-9 hardback

Transferred to digital printing 2005

Contents

Introduction and history		*page* 1
1	**Preliminaries**	5
1.1	Irreducibility and aperiodicity	6
1.2	Classification	8
1.3	Continuous time	9
1.4	Classical examples	11
1.4.1	Doeblin's condition	11
1.4.2	Birth and death process	12
1.4.3	The space homogeneous random walk on $\mathbf{Z}^m$	13
2	**General criteria**	16
2.1	Criteria involving semi-martingales	16
2.2	Criteria for countable Markov chains	26
3	**Explicit construction of Lyapounov functions**	33
3.1	Markov chains in a half-strip	33
3.1.1	Generalizations and problems	35
3.2	Random walks in $\mathbf{Z}_+^N$: main definitions and interpretation	37
3.3	Classification of random walks in $\mathbf{Z}_+^2$	39
3.4	Zero drifts	56
3.5	Jackson networks	62
3.6	Asymptotically small drifts	72
3.7	Stability and invariance principle	76
4	**Ideology of induced chains**	79
4.1	Second vector field	79
4.2	Classification of paths	82
4.3	Gluing Lyapounov functions together	86
4.4	Classification in $\mathbf{Z}_+^3$	92

5	**Random walks in two-dimensional complexes**	98
5.1	Introduction and preliminary results	98
5.2	Random walks on hedgehogs	103
5.3	Formulation of the main result	104
5.4	Quasi-deterministic process	108
5.5	Proof of the ergodicity in theorem 5.3.4	111
5.6	Proof of the transience	114
5.7	Proof of the recurrence	118
5.8	Proof of the non-ergodicity	121
5.9	Queueing applications	123
5.10	Remarks and problems	130
6	**Stability**	131
6.1	A necessary and sufficient condition for continuity	131
6.2	Continuity of stationary probabilities	137
6.3	Continuity of random walks in $\mathbf{Z}_+^N$	144
7	**Exponential convergence and analyticity**	148
7.1	Analytic Lyapounov families	148
7.2	Proof of the exponential convergence	150
7.3	General analyticity theorem	157
7.4	Proof of analyticity completed	161
7.5	Examples of analyticity	163
Bibliography		165
Index		168

Introduction and history

Introduction

This book differs essentially from the existing monographs on countable Markov chains. It intends to be, on the one hand, much *more constructive* than books similar to, for example Chung's [Chu67] and, on the other hand, much *less constructive* than some elementary monographs on queueing theory, where the emphasis is mainly put on the derivation of explicit expressions. The method of generating functions, which is to be sure the most constructive approach, is not included, since the dimension of the problems it can solve is small (in general ≤ 2). Our book could equally be called *Constructive use of Lyapounov functions method.* Here the term *constructive* is taken in the sense close to the one widely accepted in constructive mathematical physics. One can say that the objects considered have a sufficiently rich structure to be *concrete,* although the results may not always be *explicit* enough, as commonly understood. Semantically, it is permissible to say that our methods are more *qualitative constructive* than *quantitative constructive.*

The main goal of the book is to provide methods allowing a complete classification (necessary and sufficient conditions) or, in other words, allowing us to say when a Markov chain is ergodic, null recurrent or transient. Moreover, it turns out that, without doing much additional work, it is possible to study the *stability* (continuity or even analyticity) with respect to parameters, the rate of convergence to equilibrium,..., etc. by using the same Lyapounov functions.

Our primary concern with necessary *and* sufficient conditions is crucial, since in many cases it is indeed trivial to get explicit necessary *or* sufficient conditions. Another peculiarity of our approach is that we do not pursue generalizations, which could be easily done by any expert in standard classical probability theory. For example, in many places, we restrict ourselves to bounded jumps, whenever the formulation would remain unchanged in the case of unbounded jumps.

The various sections of chapter 1 give only exact definitions and some results taken from countable Markov chains that we use. To render the book accessible for the beginner, we also present section 1.4, to demonstrate the possibilities of perhaps more exact, but also more restrictive, elementary methods.

In chapter 2, we present the main classification criteria for general countable Markov chains, which are needed in the following chapters. Further far reaching martingale criteria are presented. Also we obtain some *exponential* bounds, which imply nice properties for the corresponding Markov chains.

The rest of the monograph is devoted to the so-called *deflected* random walks in $\mathbf{Z}_+^N$. The reader might wonder why random walks in $\mathbf{Z}_+^N$ are of primary interest. There are several striking reasons. First, they describe many networks of practical interest (e.g. see section 3.2) and the methods presented here could also be useful for more general networks, for instance with non-identical customers. Secondly, the problems involved not only are of probabilistic interest, but they also produce a large store of examples and, moreover, are closely connected with other branches of mathematics. In fact the classification problem for random walks in R_+^N is a probabilistic version of a well known question in functional analysis and partial differential equations: When is a multidimensional Toeplitz (or any general elliptic) operator in $\mathbf{Z}_+^N$ invertible? It also has much in common with the problem of the behaviour of diffusion processes near non-smooth boundaries of large codimension. The ideas and methods exhibited here are, in our opinion, useful for attacking problems of very different nature.

Chapter 3 gives techniques for an explicit geometrical construction of Lyapounov functions. They apply to random walks in $\mathbf{Z}_+^2$, as well as to the famous Jackson networks in $\mathbf{Z}_+^N$. The zero drift case in $\mathbf{Z}_+^2$ and *almost* zero drift one-dimensional examples of sections 3.6 and 3.7 constitute new directions of development, initiated by Lamperti [Lam60] thirty years ago. They are directly related to several works of R. Williams and others [VW85, Wil85].

The central method of *induced* chains and vector fields is presented in sections 4.1 and 4.2. In section 4.3, general results pertaining to the construction of Lyapounov functions in a uniformly bounded number of steps are given. Using these results, we obtain the complete classification in $\mathbf{Z}_+^3$.

Completely new phenomena appear in chapter 5: *scattering*, null recurrence for a positive Lebesgue measure in the parameter space, constants L and M in the simplest situation.

General criteria (some of them using Lyapounov functions), concerning conditions ensuring the continuity of stationary probabilities with respect to the parameters, are given in chapter 6.

Finally, chapter 7 offers a probabilistic criterion, again using the Lyapounov functions and Foster's theorems, for a family of Markov chains to be an *analytic Lyapounov family.* In particular, this property leads to analytic dependence on the parameters, as well as exponential convergence to equilibrium and exponential decrease of stationary probabilities.

Historical comments

Chapter 1. For the contents of this chapter we refer the reader to any standard textbooks on countable Markov chains, for example [Chu67, Kar68].

Chapter 2. The notion of *Lyapounov function* or *test function* similar to the well known Lyapounov functions for ordinary differential equations goes back to Foster [Fos53], as far as we know. Although his examples are now trivial, his ideas and criteria for ergodicity and for transience became basic for later extensions. There exist now many technical generalizations of these criteria, some of which we give in this chapter. Generalized Foster criteria for ergodicity were given in [Mal93]. In [Fil89] a new martingale proof is proposed with an important extension to random times. We have summarized and simplified all these results in theorems 2.1.1, 2.1.2 and 2.1.3. Theorems 2.2.1 and 2.2.2 extend, with new proofs, results contained in [MSZ78] and [Fos53]. Theorems 2.2.2 and 2.2.3 are the famous Foster criterion itself, with a slight modification and modern proofs. Theorem 2.2.6 generalizes some corresponding results of [Mal73] (given for a.s. uniformly bounded jumps). Theorem 2.2.8 is contained in [FMM92] and seems to be the unique constructive result allowing us to prove non-ergodicity by means of non-piecewise-linear Lyapounov functions. Theorems 2.1.1 and 2.1.10 are fundamental tools for proving all criteria we need and they also provide exponential estimates which are used in various parts of the book.

Chapter 3. Section 3.1 shows an elementary example. The results have been partially known for 20 years already. The proofs given in the book are pedagogic. Section 3.2 contains definitions taken from [MM79] and [Mal93]. Most of the theorems of sections 3.3 to 3.7 are new. The idea of using quadratic forms and functionals of quadratic forms is original, and appeared, as far as we know, for the first time in [Fay89, FB88, FMM92]. They are used in connection with the principle of almost linearity introduced in [Mal72a].

Chapter 4. Section 4.1, 4.3, 4.4 are taken, with some improvements, from [MM79]. Section 4.2 is basically contained in [Mal93].

The results of chapter 5 were first published in [FIVM91].

The content of chapters 6 and 7 is a substantial revision of the results in [MM79].

1
Preliminaries

In sections 1.1, 1.2 and 1.3 of this chapter, we briefly introduce basic notions and some results borrowed from the theory of discrete time homogeneous countable Markov chains (MC).

In section 1.4, some well known examples of MCs are given, for which a complete classification can be obtained by elementary methods: simple probabilistic arguments in 1.4.1, explicit solution of recurrent equations in 1.4.2, generating functions in 1.4.3.

It is not our intention to devote a detailed section to the fundamentals of probability theory, which are presented in a plethora of excellent textbooks. Thus, we only introduce in fact the minimal basic notions and notation useful for our purpose.

- The events are the subsets of some abstract set Ω, which belong to Σ, the σ-algebra defined on Ω.
- The couple (Ω, Σ) is a *measurable space* and the sets belonging to Σ are
  Σ-*measurable sets.*
- The triple (Ω, Σ, μ), where μ is a positive measure defined on Σ, is a *measure space.* A probability space is a measure space of total measure 1, i.e. $\mu(\Sigma) = 1$, and in this case most of the time we shall write (Ω, Σ, P).
- A Σ-measurable real-valued function f with domain Ω is called a *random variable.* More generally a *random element* φ with values in a measurable space $(X, \mathcal{B})$ is a measurable mapping of (Ω, Σ, P) into $(X, \mathcal{B})$. For $X = \mathbf{R}^N$ or $\mathbf{Z}^N$, $\mathcal{B}$ being the σ-algebra of Borel sets, we shall speak of random vectors.

1.1 Irreducibility and aperiodicity

Let $\mathcal{A}$ be a denumerable set and $\mathbf{P}$ a stochastic (transition) matrix such that

$$\mathbf{P} = (p_{\alpha\beta})_{\alpha,\beta\in\mathcal{A}}$$

and, for any $\alpha \in \mathcal{A}$, $P_\alpha = (p_{\alpha\beta})_{\beta\in\mathcal{A}}$ is a probability vector, that is

$$\sum_{\beta\in\mathcal{A}} p_{\alpha\beta} = 1 \ , \ p_{\alpha\beta} \geq 0 \, .$$

Definition 1.1.1 *The pair* $(\mathcal{A}, \mathbf{P})$ *is called a discrete time homogeneous Markov chain* (MC).

A *path* ω is any sequence

$$\omega = (\omega_0, \omega_1, \omega_2, \ldots),$$

where

$$\omega_i \in \mathcal{A} \, , \ \forall i \geq 0.$$

The *path space* $\Omega = \mathcal{A}^{\mathrm{N}}$ is the set of all paths and Σ is the standard σ-algebra generated by the cylinder sets

$$(\alpha_0 \ , \ \alpha_1, \ldots, \alpha_n) = \{\omega : \omega_i = \alpha_i \ , \ 0 \leq i \leq n\}, \ n > 0, \ \alpha_i \in \mathcal{A}.$$

Occasionally, it will be necessary to consider MC with a fixed initial distribution. Therefore we give the following

Definition 1.1.2 *We call an* MC *with initial distribution* $p_0(\alpha), \alpha \in \mathcal{A}$, $\sum_\alpha p_0(\alpha) = 1, p_0(\alpha) \geq 0$, *a probability measure* P *defined on* (Ω, Σ) *such that, for all cylinder sets* $(\alpha_0, \alpha_1, \ldots, \alpha_n)$,

$$P(\alpha_0, \alpha_1, \ldots, \alpha_n) = p_0(\alpha_0) p_{\alpha_0\alpha_1} \cdots p_{\alpha_{n-1}\alpha_n} \, . \tag{1.1}$$

The random variable $\xi_n(\omega) = \omega_n$, defined on (Ω, Σ, P) and taking its values in $\mathcal{A}$, will be called *the value of the chain at time n*, or *the position of the chain at time n*, etc. We shall simply write ξ_n, *ad libitum* and whenever unambiguous; ξ_0 is called an initial state. If there exists a sequence $\alpha_1, \alpha_2, \ldots, \alpha_{n-1}$ such that $p_{\alpha\alpha_1} p_{\alpha_1\alpha_2} \cdots p_{\alpha_{n-1}\beta} > 0$, we shall write $\alpha \rightsquigarrow \beta$.
Let us denote by $p_{\alpha\beta}^{(k)}$ the k-step transition probabilities, i.e. the elements of the matrix $\mathbf{P}^k$.

Definition 1.1.3 *The point α is called an inessential state of the MC, iff there exists a point β such that $\alpha \rightsquigarrow \beta$ but $\beta \not\rightsquigarrow \alpha$. All other states are called essential.*

It is easy to show that, for any initial state and any inessential state α, there exists a random time $N(\omega) < \infty$ such that ξ_n never equals α for $n > N(\omega)$, a.s. We shall write $\alpha \Longleftrightarrow \beta$ iff $\alpha \rightsquigarrow \beta$ and $\beta \rightsquigarrow \alpha$. The operation '$\Longleftrightarrow$' is obviously transitive. Sometimes we shall also say that α and β *communicate.*

Definition 1.1.4 *An equivalence class with respect to the operation '$\Longleftrightarrow$' is called an* essential class. *A Markov chain is called* irreducible *iff every state can be reached from any other state or, equivalently, if, and only if, $\mathcal{A}$ forms a single class of communicating states, which then are all essential.*

It is not difficult to prove that, for any initial distribution, there exists $N(\omega)$ such that all ξ_n's belong to the same essential class, for $n > N(\omega)$, almost surely (a.s.). As we shall be mainly interested in the long run behaviour of all random processes which will be encountered, from now on and for the rest of the book, *the Markov chain $(\mathcal{A}, \mathbf{P})$ will be assumed to be irreducible.*

Choose now $\alpha \in \mathcal{A}$. Let $n_1(\alpha) < n_2(\alpha) < \ldots$ be all the positive integers for which $p^{(n_i)}(\alpha, \alpha) > 0$, $i = 1, 2 \ldots$.

Definition 1.1.5 (Theorem) *Let us denote by $d(\alpha)$ the greatest common divisor of the $n_i(\alpha)$, $i \geq 1$. Then $d(\alpha)$ indeed does not depend on α and is called the* period *of the (irreducible) chain $\mathcal{A}$. If $d = 1$, the chain is called* aperiodic.

In the sequel, we shall consider only aperiodic chains, but all the theory can easily be transcribed with minor modifications to include the periodic case. In fact, it suffices to consider ξ_n at embedded instants $n = k + dm$, for some fixed k. It is also useful to keep in mind that, if for some α, $p_{\alpha\alpha} > 0$, then the chain is aperiodic. Unless otherwise stated, all the chains studied hereafter will be assumed to be *irreducible and aperiodic.*

1.2 Classification

Let $\alpha, \beta \in \mathcal{A}$. We define now, for $n \geq 1$,

$$f_n(\alpha,\beta) = \mathbf{P}(\xi_k(\omega) \neq \beta, 0 < k < n; \xi_n(\omega) = \beta / \xi_0(\omega) = \alpha) ,$$

the probability that the MC first enters into state β at time n, given that it starts from the state α. Then

$$Q(\alpha,\beta) = \sum_{n=1}^{\infty} f_n(\alpha,\beta)$$

is the probability that, starting at α, the MC ever visits β. Accordingly,

$$m_{\alpha\beta} = \sum_{n=1}^{\infty} n f_n(\alpha,\beta)$$

is the mean time of first reaching β when starting at α. Clearly $m_{\alpha\beta} = \infty$ if $Q(\alpha,\beta) < 1$.

Theorem 1.2.1 *If $Q(\alpha,\beta) = 1$ for some pair (α,β), then $Q(\alpha,\beta) = 1$ for all (α,β). Similarly, if $m_{\alpha\beta} + m_{\beta\alpha} = \infty$ for some (α,β), then $m_{\alpha\beta} + m_{\beta\alpha} = \infty$, for all (α,β) (in either instance, α and β need not be distinct).* ■

Definition 1.2.2 *An irreducible aperiodic MC is called*

(i) recurrent *if $Q(\alpha,\beta) = 1$, at least for one pair (α,β);*
(ii) non recurrent *or* transient *if $Q(\alpha,\beta) < 1$, $\forall(\alpha,\beta)$;*
(iii) positive recurrent *or* ergodic, *if $m_{\alpha\beta}+m_{\beta\alpha} < \infty$, at least for one pair (α,β);*
(iv) null recurrent *if $Q(\alpha,\beta) = 1$ and $m_{\alpha,\beta} = \infty$, at least for one pair (α,β).*
(v) non ergodic *if $m_{\alpha,\beta} = \infty$, at least for one pair (α,β).*

The purpose of the next theorems is to give other useful (equivalent) criteria for an MC to be ergodic. We consider the equation

$$\pi = \pi\mathbf{P} \text{ or, equivalently, } \pi_\beta = \sum_\alpha \pi_\alpha p_{\alpha\beta} , \tag{1.2}$$

where π is the unknown vector

$$\pi = (\pi_\alpha, \alpha \in \mathcal{A}) .$$

Theorem 1.2.3 *The limits*

$$v_\beta = \lim_{n\to\infty} p^{(n)}_{\alpha\beta} \ , \ \ \forall \beta \in \mathcal{A} \, , \tag{1.3}$$

exist and are independent of the initial state α. Futhermore, when the MC is non-ergodic, $v_\beta = 0$, $\forall\beta$.
When the MC is ergodic, then we have

$$v_\beta > 0 \ , \ \ \sum_\beta v_\beta = 1$$

and

$$v_\beta = \sum_\alpha v_\alpha p_{\alpha\beta} \ ,$$

i.e. the vector v is a probabilistic solution of (1.2). ■

Theorem 1.2.4 *The following conditions are equivalent:*

(i) *the MC is ergodic;*
(ii) *there exists a unique l^1-solution of the equation (1.2), up to a multiplicative factor;*
(iii) *there exists a unique stationary distribution $(\pi_\alpha, \alpha \in \mathcal{A})$, i.e. a solution of (1.2) such that $\pi_\alpha \geq 0$, $\sum_\alpha \pi_\alpha = 1$. In this case $\pi_\alpha > 0$, $\forall \alpha \in \mathcal{A}$,*

and

$$\pi_\alpha = \lim_{n\to\infty} p^{(n)}_{\gamma\alpha} \ , \ \forall \gamma \in \mathcal{A} \, . \tag{1.4}$$

■

Theorem 1.2.5 *For an ergodic MC, the invariant distribution is given by*

$$\pi_\alpha = \frac{1}{m_{\alpha\alpha}} \ , \quad \forall \alpha \in \mathcal{A} \, . \tag{1.5}$$

■

1.3 Continuous time

Many examples seem more natural in continuous time. Later on we introduce the necessary notation to the extent we need. But we want to stress immediately that all results concerning the classification in discrete time are automatically transposed into continuous time and vice versa.

There are two main definitions of a continuous time homogeneous countable MC (the set of states is still denoted by $\mathcal{A}$). In both cases, the *intensity matrix* $H = (\lambda_{\alpha\beta})$ is given by

$$\begin{cases} \lambda_{\alpha\beta} \geq 0,\ \alpha \neq \beta\,, \\ \lambda_{\alpha\alpha} = -\sum\limits_{\beta:\beta\neq\alpha} \lambda_{\alpha\beta}\,. \end{cases} \tag{1.6}$$

For the examples we shall consider, it suffices to assume the existence of a constant $C > 0$ such that, for all α, β,

$$|\lambda_{\alpha\alpha}| < C\,. \tag{1.7}$$

Then the matrix

$$\| p_{\alpha\beta}(t) \| = P(t) = e^{Ht} = \sum_{n=0}^{\infty} H^n \frac{t^n}{n!}\,,\ t \geq 0\,, \tag{1.8}$$

is defined by the convergent series (1.8) .

Definition 1.3.1 *The MC ξ_t, with initial distribution $p_\alpha(0)$, is defined by the following finite-dimensional distributions, for all $0 < t_1 < \ldots < t_n$:*

$$P(\xi_0 = \alpha_0, \ldots, \xi_{t_n} = \alpha_n) = p_{\alpha_0}(0) p_{\alpha_0\alpha_1}(t_1) \ldots p_{\alpha_{n-1}\alpha_n}(t_n). \tag{1.9}$$

This definition does not depend on the choice of the probability space. The next one uses a concrete choice. We define Ω to be the set of right-continuous piecewise constant mappings $\omega : [0, \infty) \to \mathcal{A}$, i.e. ω is given by a sequence $(\alpha_0, 0), (\alpha_1, \tau_1), \ldots$, such that

$$\omega(t) = \alpha_i,\ t \in [\tau_i, \tau_{i+1}), \tau_0 = 0\,,$$

where $\tau_1, \tau_2, \ldots,$ are the *jump times*. The measure on Ω, corresponding to the MC (using the standard canonical σ-algebra, see for example [Chu67, GS74]), is defined by the following conditions:

(i) Given $\alpha_0, \alpha_1, \ldots,$ the random variables $\tau_{i+1} - \tau_i$ are mutually independent and have an exponential distribution with parameters $-\lambda_{\alpha_i\alpha_i}$;

(ii) $\alpha_0, \alpha_1, \ldots, \alpha_n, \ldots,$ are distributed as an *embedded* discrete time homogeneous MC, with parameters

$$p_{\alpha\beta} = \frac{\lambda_{\alpha\beta}}{(-\lambda_{\alpha\alpha})}. \tag{1.10}$$

It is easy to show that this measure leads to the finite-dimensional probabilities (1.9). Below we always assume the embedded chain to be irreducible and aperiodic. The classification of such continuous time MCs reduces to the classification of the corresponding imbedded chain.

1.4 Classical examples

This book is intended to provide general methods to classify Markov chains in terms of ergodicity, null recurrence or transience. In some (rare) cases, it is possible to get a complete answer from some elementary consideration or by finding explicit tractable expressions for $Q_{\alpha\beta}, m_{\alpha\beta}$ or π_α defined in preceding sections.

1.4.1 Doeblin's condition

Take an MC $(\mathcal{A}, \mathbf{P})$ satisfying the following simple condition, due to Doeblin: There exist a finite set $\mathcal{A}_0$, an integer $j > 0$ and a real number $\epsilon > 0$ such that, for all $\alpha \in \mathcal{A}$,

$$p^{(j)}(\alpha, \mathcal{A}_0) > \epsilon ,$$

where

$$p^{(j)}(\alpha, \mathcal{A}_0) = \sum_{\beta \in \mathcal{A}_0} p^{(j)}(\alpha, \beta).$$

It is immediate from theorem 1.2.4 that such Markov chains are ergodic. Indeed, since

$$p^{(k)}(\alpha, \mathcal{A}_0) = \sum_{\beta} p_{\alpha\beta}^{(k-j)} p^j(\beta, \mathcal{A}_0) > \epsilon \ , \quad \forall k \geq j \ , \quad \forall \alpha \in \mathcal{A} \, ,$$

we have

$$\pi_{\mathcal{A}_0} \stackrel{\text{def}}{=} \sum_{\beta \in \mathcal{A}_0} \pi_\beta = \sum_{\beta \in \mathcal{A}_0} \lim_{n \to \infty} p_{\alpha\beta}^{(n)} = \lim_{n \to \infty} \sum_{\beta} p_{\alpha\beta}^{(n)} = \lim_{n \to \infty} p^{(n)}(\alpha, \mathcal{A}_0) \geq \epsilon.$$

A direct argument could also be used : at least one point $\alpha_0 \in \mathcal{A}_0$ is entered infinitely often, with a finite mean hitting time. Note that all irreducible MCs with a finite number of states do satisfy Doeblin's condition and, therefore, are ergodic.

1.4.2 Birth and death process

A birth and death process is an MC with state space $\mathbf{Z}_+$, which has the following transition probabilities:

$$p_{i,i+1} = p_i \ , \ p_{i,i-1} = q_i \ ,$$

$$p_i + q_i = 1 \ , \ i \geq 1 \ , \ p_0 = 1, q_0 = 0 \ .$$

Let us put

$$\begin{cases} \pi_0 = 1 \ , \ \pi_i = \dfrac{p_1 p_2 \cdots p_{i-1}}{q_1 q_2 \cdots q_i}, \ i \geq 1 \ , \\ \\ A = \displaystyle\sum_{i=0}^{\infty} \pi_i \ , \ B = \sum_{i=0}^{\infty} \frac{1}{p_i \pi_i}. \end{cases} \tag{1.11}$$

Theorem 1.4.1 *A birth and death process is*

(i) *ergodic* if, and only if, $A < \infty$;
(ii) *null recurrent* if, and only if, $A = B = \infty$;
(iii) *transient* if, and only if $B < \infty$.

Proof: Consider the system of equations for the stationary distribution $\{\pi_i\}$,

$$\pi_i = \pi_{i-1} \, p_{i-1} + \pi_{i+1} \, q_{i+1}. \tag{1.12}$$

It is clear that they have a nonzero unique solution, given by (1.11) up to a constant factor. So (i) follows from theorem 1.2.3. Equations for the probabilities y_i of ever reaching 0, starting from i, are

$$y_i = p_i y_{i+1} + q_i y_{i-1}, \ i \geq 1. \tag{1.13}$$

It can be easily verified that $y_n^{(0)} \equiv 1$ is a solution of (1.13), another solution being given by

$$y_0^{(1)} = 0, \ y_n^{(1)} = \sum_{i=0}^{n-1} \frac{1}{p_i \pi_i}, i \geq 1 \ .$$

Hence, the general solution has the form

$$y_n = C_0 \, y_n^{(0)} + C_1 \, y_n^{(1)} \ .$$

If $B = \infty$ then $y_n^{(1)} \to \infty$, as $n \to \infty$, and the only probabilistic solution is $y_n^{(0)}$, so that the MC is recurrent. If $B < \infty$, there is another probabilistic solution

$$y_n = 1 - \frac{1}{B} \, y_n^{(1)} \ ,$$

where *probabilistic solution* means $y_0 = 1$ and $0 < y_n \leq 1, n \geq 1$. Let us note now that any probabilistic solution y_i satisfies the equations

$$\sum_{j=0}^{\infty} \tilde{p}_{ij}\, y_j = y_i \ , \quad i = 0, 1, \ldots \tag{1.14}$$

where $\tilde{p}_{ij} = p_{ij}$, for $i > 0$, $\tilde{p}_{0i} = \delta_{0i}$.
The transition probabilities $\tilde{p}_{ij}$ define a new (reducible) MC having an absorbing state at 0.
After iterating (1.14), we get

$$\sum_{j=0}^{\infty} \tilde{p}_{ij}^{(n)} y_j = y_i$$

and, since $y_0 = 1$,

$$\tilde{p}_{i0}^{(n)} \leq y_i \ .$$

But $\lim \tilde{p}_{i0}^{(n)}$ is the probability, for the initial MC, of being absorbed into 0, starting from i. So, if there exists a probabilistic solution with $y_i < 1$ for some i, then the MC is transient. ∎

1.4.3 The space homogeneous random walk on $\mathbf{Z}^m$

We shall denote by $\mathbf{Z}^m$ the lattice of all integer-valued vectors in the space $\mathbf{R}^m$. The position of the random walk at time n is defined by a random vector $\xi_n \in \mathbf{Z}^m$, such that

$$\begin{cases} \xi_n = \alpha + \eta_1 + \eta_2 + \ldots \eta_n, \ n \geq 1, \\ \xi_0 = \alpha \ , \end{cases}$$

where $\alpha \in \mathbf{Z}^m$ is a deterministic vector giving the original position of the particle at time 0, and $\eta_k, k \geq 1$, are i.i.d. random vectors with range $\mathbf{Z}^m$.
It is immediate that ξ_n is a discrete time MC, which is, moreover, spatially homogeneous, in the sense that its transition probabilities satisfy

$$p_{\alpha\beta} = g(\beta - \alpha) \ , \quad \forall \alpha, \beta \in \mathbf{Z}^m \ ,$$

where we have put

$$g(\gamma) = P(\eta_k = \gamma) \ , \quad \gamma \in \mathbf{Z}^m \ .$$

We quote only the main results, referring the reader to [GS74] for a detailed treatment. The classical way of analysing the random walk

relies on the method of characteristic functions.
Let

$$F(u) = E[e^{i(u,\eta_1)}] ,$$

where $u = (u_1, u_2, \ldots, u_m) \in \mathbf{R}^m$.
For an MC to be irreducible, a well known necessary and sufficient condition (see [GS74]) is that

$$F(u) \neq 1, \text{ for } u \neq 2\pi\alpha, \ \alpha \in \mathbf{Z}^m .$$

First, note that the random walk is never ergodic, as emerges easily from a translation invariance argument. Secondly, if $E(\eta_1) \neq 0$, then the random walk is transient, as can be seen by using the law of large numbers.

Theorem 1.4.2 *For $m \geq 3$, the random walk is always non-recurrent. It is also non-recurrent if $E(\eta_1) \neq 0$. If $E(\eta_1) = 0$, then the random walk is recurrent for $m = 1$. If $E(\eta_1) = 0$, $E(\eta_1^2) < \infty$, then the random walk is recurrent for $m = 2$.*

Proof : Only the case $E(\eta_1) = 0$ needs to be considered. We shall use the following general criterion for the recurrence of a MC (see for instance [Kar68, GS74].

Theorem 1.4.3 *An irreducible aperiodic MC is recurrent if, and only if,*

$$\sum_n p_{ii}^{(n)} = \infty, \quad \textit{for some } i, \textit{ and then for all } i. \qquad \blacksquare$$

We do not use the criterion of theorem 1.4.3 in the sequel. We simply quote that, in the convergent case,

$$\sum_{n=0}^{\infty} p_{ii}^{(n)} = \frac{1}{1 - Q(i,i)} .$$

Hence, for the random walk to be recurrent, it is necessary and sufficient to have

$$G = \sum_{n \geq 0} g^{(n)}(0) = \infty ,$$

where $g^{(n)}(\cdot)$ denotes the n-th iterate of the probability distribution function $g(\cdot)$ defined above. The following equality holds:

$$G = \lim_{z \uparrow 1} R(z) , \ 0 < z < 1 ,$$

where

$$R(z) = \sum_{n\geq 0} g^{(n)}(0)z^n = \frac{1}{(2\pi)^m} \sum_{n\geq 0} \int_D F^n(u)z^n du = \frac{1}{(2\pi)^m} \int_D \frac{du}{1 - zF(u)} ,$$

and $D = \{u : | u_i | \leq \pi, i = 1, \ldots, m\}$. Since z is real, we get

$$G = \lim_{z\uparrow 1} \frac{1}{(2\pi)^m} \int_D \mathbf{Re}[1 - zF(u)]^{-1} du .$$

This shows that the boundedness of G is equivalent to the convergence of the above integral and the theorem is proved. ■

2
General criteria

In this chapter we present and prove several general criteria, which are constantly used throughout the book. *En passant*, it is worth mentioning that:

(i) some other criteria exist [Szp90], which in fact we could not effectively use for our constructive problems, so that we shall not discuss them;

(ii) although martingale or Lyapounov function ideology is indispensable and could be perceived as fundamental for such criteria, we realize that some deeper meta-theory for producing such criteria might well exist too.

2.1 Criteria involving semi-martingales

Let $(\Omega, \mathcal{F}, P)$ be a given probability space and $\{\mathcal{F}_n, n \geq 0\}$ an increasing family of σ-algebras $\mathcal{F}_0 \subset \mathcal{F}_1 \subset \ldots \subset \mathcal{F}_n \subset \ldots \subset \mathcal{F}$. Let $\{S_i, i \geq 0\}$ be a sequence of real non negative random variables, such that S_i is $\mathcal{F}_i$-measurable, $\forall i \geq 0$. Moreover, S_0 will be taken constant. Denote by τ the $\mathcal{F}_n$-stopping time representing the epoch of the first entry into $[0, C]$, i.e. $\tau(\omega) = \inf\{n \geq 1 : S_n(\omega) \leq C\}$. Introduce the stopped sequence $\tilde{S}_n = S_{n\wedge\tau}$, where

$$n \wedge \tau = \begin{cases} n, & \text{if } n \leq \tau , \\ \tau, & \text{if } n > \tau . \end{cases}$$

We also use the classical notation for the indicator function

$$\mathbf{1}_{\mathcal{A}} = \begin{cases} 1 , & \text{if } \mathcal{A} \text{ is true,} \\ 0 , & \text{otherwise.} \end{cases}$$

Theorem 2.1.1 *Assume that* $S_0 > C$ *and, for some* $\epsilon > 0$ *and all* $n \geq 0$,

$$E(\tilde{S}_{n+1}/\mathcal{F}_n) \leq \tilde{S}_n - \epsilon \mathbf{1}_{\{\tau>n\}} \text{ a.s.} \tag{2.1}$$

Then

$$E(\tau) < \frac{S_0}{\epsilon} < \infty\,. \tag{2.2}$$

Proof : Taking expectation in (2.1) yields

$$E(\tilde{S}_{n+1} - \tilde{S}_n) \leq -\epsilon P(\tau > n)$$

and, by summing over n and taking into account $\tilde{S}_i \geq 0$,

$$0 \leq E(\tilde{S}_{n+1}) \leq -\epsilon \sum_{i=0}^{n} P(\tau > i) + S_0\,,$$

which implies

$$E(\tau) = \lim_{n\to\infty} \sum_{i=0}^{n} P(\tau > i) < \frac{S_0}{\epsilon} < \infty\,.$$

This proves the theorem. ■

Now we shall formulate and prove a theorem which generalizes theorem 2.1 and will be an important instrument in the investigation of the ergodicity of random walks in $\mathbf{Z}_+^n$. Let $\{N_i, i \geq 1\}$ be an increasing sequence of stopping times of S_n, i.e. $\{N_i = n\} \in \mathcal{F}_n$, for all n and i, and such that $N_0 = 0$, $N_i - N_{i-1} \geq 1$, a.s. $\forall i \geq 1$. Introduce $Y_0 = S_0$, $Y_i = S_{N_i}$, $i \geq 1$, the stopping time

$$\sigma = \inf\{i \geq 1 : Y_i \leq C\}\,,$$

and the stopped sequences $\tilde{Y}_i = Y_{i\wedge\sigma}$, $\tilde{N}_i = N_{i\wedge\sigma}$, $i \geq 1$.

Theorem 2.1.2 *Assume* $S_0 > C$ *and, for some* $\epsilon > 0$ *and all* $n \geq 0$,

$$E(\tilde{Y}_{n+1}/\mathcal{F}_{N_n}) \leq \tilde{Y}_n - \epsilon\, E(\tilde{N}_{n+1} - \tilde{N}_n/\mathcal{F}_{\tilde{N}_n}), \text{ a.s.} \tag{2.3}$$

Then

$$E(\tau) \leq \frac{S_0}{\epsilon}\,. \tag{2.4}$$

Proof : It follows from (2.3) that

$$E(\tilde{Y}_i - \tilde{Y}_{i-1}) = E[E(\tilde{Y}_i - \tilde{Y}_{i-1}/\mathcal{F}_{\tilde{N}_{i-1}})]$$

$$\leq -\epsilon E(\tilde{N}_i - \tilde{N}_{i-1})\,.$$

Consequently,

$$E(\tilde{Y}_n) \leq -\epsilon \sum_{i=1}^{n} [E(\tilde{N}_i) - E(\tilde{N}_{i-1})] + S_0$$

$$= -\epsilon\, E(\tilde{N}_n) + S_0 \,. \tag{2.5}$$

Since $\tilde{Y}_n \geq 0$ a.s., we obtain from (2.5)

$$E(\tilde{N}_n) \leq \frac{S_0}{\epsilon} \;,\quad \forall n \geq 1 \,. \tag{2.6}$$

Thus, as follows from theorem 2.1.1, $E(\sigma) < \infty$. Since $\tilde{N}_i$ is pointwise increasing with respect to i, we get, using the monotone convergence theorem and (2.6),

$$E(N_\sigma) = E(\lim_{n\to\infty} N_{n\wedge\sigma}) = \lim_{n\to\infty} E(N_{n\wedge\sigma}) \leq \frac{S_0}{\epsilon} \,. \tag{2.7}$$

Also, for any sample path $S_0, S_1, \ldots, S_i, \ldots$, it is immediate that

$$\tau \leq N_\sigma \;\text{ a.s.}$$

Consequently

$$E(\tau) \leq E(N_\sigma) < \frac{S_0}{\epsilon} < \infty \,, \tag{2.8}$$

which proves (2.4) and the theorem. ■

Theorem 2.1.3 *Suppose $S_0 > C$ and, for $n \geq 1$ and some positive real M,*

$$E(\tilde{S}_n / \mathcal{F}_{n-1}) \geq \tilde{S}_{n-1} \;,\; a.s. \,, \tag{2.9}$$

$$E(|\, \tilde{S}_n - \tilde{S}_{n-1} \,| / \mathcal{F}_{n-1}) \leq M \;\; a.s. \tag{2.10}$$

Then $E(\tau) = \infty$. (Here the S_n 's are not necessarily positive.)

Proof : For all $k \geq 1$, we get from (2.10)

$$E(|\, \tilde{S}_k - \tilde{S}_{k-1} \,|) = E[E(|\, \tilde{S}_k - \tilde{S}_{k-1} \,| / \mathcal{F}_{k-1})] \leq MP(\tau > k-1) \,.$$

Thus, for any n, l, such that $1 \leq l \leq n$,

$$E(|\, \tilde{S}_n - \tilde{S}_l \,|) = E[|\sum_{k=l+1}^{n} (\tilde{S}_k - \tilde{S}_{k-1})\,|] \leq \sum_{k=l+1}^{n} E(|\, \tilde{S}_k - \tilde{S}_{k-1} \,|)$$

$$\leq M \sum_{k=l+1}^{n} P(\tau \geq k) , \tag{2.11}$$

whence, immediately,

$$E(| \tilde{S}_n |) \leq M \sum_{k=0}^{n} P(\tau > k) + S_0. \tag{2.12}$$

Assume $E(\tau) < \infty$. Then, from (2.11), (2.12) and Cauchy's criterion, it follows that $\tilde{S}_n$ is a submartingale converging almost surely (a.s) in L_1. [The convergence a.s. is here obvious since, by hypothesis, $P(\tau < \infty) = 1$ and thus $\tilde{S}_n = S_{n \wedge \tau} \overset{a.s.}{\rightarrow} S_\tau$]. Thus we have

$$E(S_\tau) = \lim_{n \to \infty} E(\tilde{S}_n) \geq E(S_0) .$$

But, from the very definition of $\tau, E(S_\tau) \leq C$, which yields a contradiction. Hence $E(\tau) = \infty$ and the proof of theorem 2.3 is concluded. ∎

Theorem 2.1.4 *Let $\{H_n, n \geq 0\}$ be a martingale belonging to L^α, $1 < \alpha \leq 2$, where $H_0 = 0$ and $\{B_n, n \geq 0\}$ denotes the increasing process associated to Doob's decomposition of $|H_n|^\alpha = U_n + B_n$, U_n being a martingale. Then*

(i) *the martingale $\{H_n, n \geq 0\}$ converges almost surely in L^α to a finite limit on the event $\{B_\infty < \infty\}$;*
(ii) *if $E(B_\infty) < \infty$, then the martingale $\{H_n, n \geq 0\}$ converges in L^α.*

Moreover, when $\alpha = 2$, $E(\sup_{n \geq 0} H_n^2) \leq 4E(B_\infty)$. ∎

For $\alpha = 2$, this theorem appears for instance in Neveu [Nev72]. The extension to the case $1 < \alpha \leq 2$ is obtained, first, by using the following classical inequality, valid for any positive submartingale $\{X_n, n \geq 0\}$ and any $p > 1$,

$$\| \sup_n X_n \|_p \leq \frac{p}{p-1} \sup_n \| X_n \|_p ,$$

and, secondly, by introducing an estimate analogous to the one derived from (2.17) in the forthcoming Lemma 2.1.6. ∎

Theorem 2.1.5 *If, for all $n \geq 1$ and α, $1 < \alpha \leq 2$,*

$$E[\tilde{S}_{n+1} - \tilde{S}_n / \mathcal{F}_n] \leq 0 \ \ a.s. , \tag{2.13}$$

$$E[| \tilde{S}_{n+1} - \tilde{S}_n |^\alpha / \mathcal{F}_n] \leq M \ \ a.s. , \tag{2.14}$$

$$E(\tau) < \infty ,$$

then

$$\tilde{S}_n^\alpha \xrightarrow{L^1} S_\tau^\alpha . \tag{2.15}$$

Before proving theorem 2.1.5, let us formulate the following lemma which is of independent interest.

Lemma 2.1.6 *If the conditions (2.13), (2.14) hold and $E(\tau) < \infty$, then*

$$\sup_n E(\tilde{S}_n^\alpha) < \infty \ , \forall \alpha, \ 1 \leq \alpha \leq 2 . \tag{2.16}$$

Proof : Define

$$\Delta \tilde{S}_n = \tilde{S}_{n+1} - \tilde{S}_n .$$

The following estimate applies, from Taylor's formula:

$$\tilde{S}_{n+1}^\alpha - \tilde{S}_n^\alpha = \alpha \Delta \tilde{S}_n (\tilde{S}_n + \theta_n \Delta \tilde{S}_n)^{\alpha - 1} , \tag{2.17}$$

where $0 < \theta_n < 1$, $\forall n \geq 0$. The right-hand side member of (2.17) can now be rewritten as

$$\alpha \tilde{S}_n^{\alpha-1} \Delta \tilde{S}_n + \alpha \tilde{S}_n^{\alpha-1} \Delta \tilde{S}_n \left[(1 + \frac{\theta_n \Delta \tilde{S}_n}{\tilde{S}_n})^{\alpha-1} - 1 \right]$$
$$\leq \alpha \tilde{S}_n^{\alpha-1} \Delta \tilde{S}_n + \alpha \mid \Delta \tilde{S}_n \mid^\alpha ,$$

where we have used the elementary inequalities

$$\mid 1 + v \mid^q \leq 1 + v^q \ , \ \mid 1 - v \mid^q \geq 1 - v^q \ , \ \forall q, \ 0 \leq q \leq 1 \ , \ \forall v \geq 0 .$$

Thus taking conditional expectation in (2.17) and using (2.13) and (2.14), we get

$$E[\tilde{S}_{n+1}^\alpha - \tilde{S}_n^\alpha / \mathcal{F}_n] \leq \alpha M \ \mathbf{1}_{\{\tau > n\}} \ \text{ a.s.} , \tag{2.18}$$

and, hence,

$$E(\tilde{S}_{n+1}^\alpha) \leq \alpha M \sum_{k=0}^{n} P(\tau > k) + S_0^\alpha \leq \alpha M E(\tau) + S_0^\alpha .$$

The finiteness of $E(\tau)$ yields (2.16) and lemma 2.1.6 is proved. ■

Proof of theorem 2.1.5 $\{\tilde{S}_n\}$ is a positive finite supermartingale. Therefore, using Doob's decomposition, we have $\tilde{S}_n = M_n - A_n$, where M_n is a positive martingale and A_n is an increasing predictable sequence. Let us prove in fact that

$$M_n = M_{n \wedge \tau} \ , \ A_n = A_{n \wedge \tau} ,$$

i.e. M_n and A_n are stopped sequences with respect to τ.
Now, from the very definition of $\tilde{S}_n$, we have on $\{\tau \leq n\}$,

$$M_n - A_n = M_{n+1} - A_{n+1} = S_\tau,$$

or, equivalently,

$$(M_n - M_{n+1})\mathbf{1}_{\{\tau \leq n\}} = (A_n - A_{n+1})\mathbf{1}_{\{\tau \leq n\}}. \tag{2.19}$$

It follows from (2.19) that

$$\begin{aligned} E((A_n - A_{n+1})\mathbf{1}_{\{\tau \leq n\}}/\mathcal{F}_n) &= E((M_n - M_{n+1})\mathbf{1}_{\{\tau \leq n\}}/\mathcal{F}_n) \\ &= \mathbf{1}_{\{\tau \leq n\}} E((M_n - M_{n+1})/\mathcal{F}_n) = 0 . \end{aligned}$$

But $(A_n - A_{n+1})\mathbf{1}_{\{\tau \leq n\}}$ is measurable with respect to $\mathcal{F}_n$, so that

$$(A_n - A_{n+1})\mathbf{1}_{\{\tau \leq n\}} = 0 \quad \text{a.s.}$$

It follows also from (2.19) that $(M_n - M_{n+1})\mathbf{1}_{\{\tau \leq n\}} = 0$ a.s. Thus M_n and A_n are stopped sequences, as asserted above. The next step consists in showing the uniform boundedness of the sequences $\mid A_{n+1} - A_n \mid^\alpha$ and $E(\mid M_{n+1} - M_n \mid^\alpha /\mathcal{F}_n)$. We know that

$$A_{n+1} - A_n = E(\tilde{S}_n - \tilde{S}_{n+1}/\mathcal{F}_n), \text{ a.s.}$$

Therefore, using Jensen's inequality for $\alpha \in [1, +\infty]$, we have

$$\mid A_{n+1} - A_n \mid^\alpha \leq E(\mid S_n - S_{n+1} \mid^\alpha /\mathcal{F}_n) \leq M \text{ a.s. },$$

so that

$$A_{n+1} - A_n \leq M^{1/\alpha} \text{ a.s.} \tag{2.20}$$

Since

$$M_{n+1} - M_n = \tilde{S}_{n+1} - \tilde{S}_n + A_{n+1} - A_n ,$$

the triangular inequality for the L^α-norm and (2.20) yields

$$\begin{aligned} (E(\mid M_{n+1} - M_n \mid^\alpha /\mathcal{F}_n))^{1/\alpha} &\leq (E(\mid \tilde{S}_{n+1} - \tilde{S}_n \mid^\alpha /\mathcal{F}_n))^{1/\alpha} \\ &\quad + A_{n+1} - A_n \leq 2M^{1/\alpha}, \end{aligned}$$

whence

$$E(\mid M_{n+1} - M_n \mid^\alpha /\mathcal{F}_n) \leq 2^\alpha M . \tag{2.21}$$

Applying now lemma 2.1.6 to the martingale M_n, we have

$$\sup_n E(M_n^\alpha) < \infty. \tag{2.22}$$

Then, using the uniform boundedness of $E(M_n^\alpha)$ and theorem 2.1.4, it follows that

$$M_n \stackrel{L^\alpha}{\to} M_\tau .$$

Since A_n is an increasing process and $0 \leq A_n \leq M_n$ a.s for all n, Lebesgue's dominated convergence theorem ensures that

$$A_n \stackrel{L^\alpha}{\to} A_\tau .$$

Finally, as $\tilde{S}_n = M_n - A_n$, the supermartingale $\tilde{S}_n$ converges in L^α. Theorem 2.1.5 is proved. ∎

Let us assume now that the S_i 's defined above are not necessarily non-negative and introduce the following random variables:

$$y_{k+1} = S_{k+1} - S_k \ , \ \tilde{y}_k = y_k \mathbf{1}_{\{y_k > b\}} \ ,$$

where b is some given constant. Usually the y_k's will be called the *the jumps* of the process $\{S_n\}$.

Theorem 2.1.7 *If there exist a constant b and positive numbers ϵ, l, such that*

$$E(\tilde{y}_{k+1}/\mathcal{F}_k) \leq -\epsilon \ a.s. \ , \tag{2.23}$$

$$y_{k+1} = S_{k+1} - S_k < l \ a.s. \ , \tag{2.24}$$

then, for any $\delta_1 < \epsilon$, there also exist constants $D = D(S_0)$ and $\delta_2 > 0$, such that, for any $n \geq 0$,

$$P(S_n > -\delta_1 n) < Ce^{-\delta_2 n} . \tag{2.25}$$

Proof : First, we note that, if (2.23) is satisfied at all, then necessarily $b \leq 0$. Secondly, for all $b \leq 0$,

$$S_n = \sum_{i=1}^{n} y_i + S_0 \ \leq \ \sum_{i=1}^{n} \tilde{y}_i + S_0 = \tilde{S}_n \ .$$

In this case

$$| \ \tilde{y}_i \ | \leq \max(-b, l) \stackrel{\text{def}}{=} d$$

and

$$P(S_n > -\delta_1 n) \leq P(\tilde{S}_n > -\delta_1 n) \ .$$

This simple remark allows us to reduce the case of jumps bounded from above (but not necessarily from below) to the simpler one, when the jumps are *bounded in absolute value.* Thus it will be assumed throughout the proof that

$$|\, y_i \,| \leq d < \infty \;\;,\;\; \forall i \geq 0\,,$$

and we shall use (2.23) without the *tilde* symbol. From Chebyshev's inequality, we have

$$\begin{aligned} P(S_n > 0) &= P\left(\sum_{k=1}^{n} y_k > -S_0\right) \qquad (2.26)\\ &= P(e^{h\sum_{k=1}^n y_k} > e^{-hS_0}) \leq e^{hS_0} E[e^{h\sum_{k=1}^n y_k}]\,, \end{aligned}$$

for any $h > 0$.

Choosing $0 < h < \frac{1}{d}$, we have

$$e^{hy_k} < 1 + hy_k + \frac{3}{2}(hy_k)^2\,,$$

which follows from the simple inequality

$$e^x < 1 + x + \frac{3x^2}{2} \;\;,\;\; |\, x \,| < 1\,.$$

Hence, by (2.23),

$$\begin{aligned} E[e^{hy_k}/\mathcal{F}_{k-1}] &< E[1 + hy_k + \frac{3}{2}(hy_k)^2/\mathcal{F}_{k-1}]\\ &\leq 1 - h\epsilon + \frac{3h^2d^2}{2} \;\;,\;\; \text{a.s.} \end{aligned}$$

Therefore, taking h sufficiently small, we obtain, for some $\delta > 0$ and all $k \geq 1$,

$$E[e^{hy_k}/\mathcal{F}_{k-1}] < e^{-\delta} \;,\;\; \text{a.s.} \qquad (2.27)$$

Hence, from (2.27)

$$\begin{aligned} E[e^{h\sum_{k=1}^n y_k}] &= E[\prod_{k=1}^{n} e^{hy_k}] = E[E(\prod_{k=1}^{n} e^{hy_k}/\mathcal{F}_{n-1})]\\ &= E[\prod_{k=1}^{n-1} e^{hy_k} E(e^{hy_n}/\mathcal{F}_{n-1})] \; < e^{-\delta} E[\prod_{k=1}^{n-1} e^{hy_k}]\,, \end{aligned}$$

which yields immediately

$$E[e^{h\sum_{k=1}^n y_k}] < e^{-n\delta}\,.$$

After setting $K = e^{hS_0}$, we obtain

$$P(S_n > 0) < Ke^{-\delta n} . \tag{2.28}$$

Let $Z_n = S_n + n\delta_1$, for any fixed $\delta_1 < \epsilon$, and $\epsilon_1 = \epsilon - \delta_1$. Then

$$E[Z_n/\mathcal{F}_{n-1}] - Z_{n-1} = E[S_n - S_{n-1}/\mathcal{F}_{n-1}] + \delta_1 \le -\epsilon + \delta_1 = -\epsilon_1 < 0 . \tag{2.29}$$

It follows from (2.29) that, for some $D, \delta_2 > 0$,

$$P(Z_n > 0) < De^{-n\delta_2} \quad , \ \forall n \ge 0 ,$$

whence

$$P(S_n > -\delta_1 n) = P(Z_n > 0) < De^{-n\delta_2} .$$

The proof of theorem 2.1.7 is concluded. ∎

The following theorem is a strenghtening of theorem 2.1.7 in the case of bounded jumps.

Theorem 2.1.8 *Let $\{N_i, i \ge 1\}$ be a strictly increasing sequence of $\mathcal{F}_n$-stopping-times, i.e. $\{N_i = n\}$ is $\mathcal{F}_n$-measurable and $N_0 = 0$. If for some $d, r, \epsilon > 0$ and all $i \ge 0$, the inequalities*

$$\begin{cases} |S_i - S_{i-1}| < d , \\ 1 \le N_i - N_{i-1} \le r , \\ E(S_{N_i}/\mathcal{F}_{N_{i-1}}) \le S_{N_{i-1}} - \epsilon , \end{cases} \tag{2.30}$$

hold with probability 1, then for any $\delta_1 < \epsilon$, there exist constants $D = D(S_0)$ and $\delta > 0$, such that, for all $n \ge 0$,

$$P(S_n > -\delta_1 n) < De^{-\delta n} . \tag{2.31}$$

Proof : Let $Y_i = S_{N_i}$, $Y_0 = S_0$. The sequence $\{Y_i \ , \ i \ge 0\}$ satisfies the assumptions of theorem 2.1.7. Therefore, for any $\delta_1 < \epsilon$, there exist $C_1, \delta_2 > 0$, such that, for all $i \ge 0$,

$$P(Y_i > -\delta_1 i) < C_1 e^{-\delta_2 i} . \tag{2.32}$$

It follows from (2.32) that there also exist $C_2, \delta_3 > 0$ such that, for all $i \ge 0$,

$$P(Y_i > -\delta_1 i - dr) < C_2 e^{-\delta_3 i} . \tag{2.33}$$

Consider the event $A_n = \{S_n > -\delta_1 n\}$. The first two conditions of (2.30) yield

$$A_n \subset \bigcup_{m=[n/r]}^{n} \{Y_m > -\delta_1 m - dr\} .$$

Consequently, taking (2.33) into account, we obtain

$$P(S_n > -\delta_1 n) = P(A_n) \leq \sum_{m=[n/r]}^{n} P(Y_m > -\delta_1 m - dr)$$

$$\leq C_2 \sum_{m=[n/r]}^{n} e^{-\delta_3 m} ,$$

which in turn implies that there exist D and δ such that

$$P(S_n > -\delta_1 n) < De^{-\delta n} \quad , \quad \forall n \geq 0 .$$

This proves the theorem. ■

Theorem 2.1.9 *Let N_i, d, r be as in theorem 2.1.8; assume that, for some $\epsilon > 0$ and all $i \geq 0$,*

$$E[S_{N_i}/\mathcal{F}_{N_{i-1}}] \geq S_{N_{i-1}} + \epsilon \quad , \quad \text{a.s.} \tag{2.34}$$

and

$$S_0 > C + dr .$$

Then $P(\tau = \infty) > 0$.

Proof : It is sufficient to prove that, for some m, there exists $\gamma > 0$, such that

$$q = P(\bigcap_{k=m}^{\infty} \{S_k > C\}) > \gamma , \tag{2.35}$$

for $S_0 > C + dr$. Clearly, we have

$$q = 1 - P(\bigcup_{k=m}^{\infty} \{S_k \leq C\}) > 1 - \sum_{k=m}^{\infty} P(S_k \leq C) . \tag{2.36}$$

Proceeding along the same lines as in theorems 2.1.7 and 2.1.8, but reversing the inequalities, we prove the existence of $\delta > 0$, and a, such that

$$P(S_k \leq C) < ae^{-\delta k} \quad , \quad \forall k \geq 0 .$$

Since the series $\sum_{k=1}^{\infty} e^{-\delta k}$ is convergent, there exist $\gamma > 0$ and m, such that

$$\sum_{k=m}^{\infty} P(S_k \leq C) < 1 - \gamma . \tag{2.37}$$

The theorem now follows from the inequalities (2.35), (2.36), (2.37). ■

We propose now a result which is, in its nature, similar to theorem 2.1.9, but assumes merely the lower boundedness of jumps.

Theorem 2.1.10 *If there exist a constant b and positive numbers ϵ, l such that*

$$y_{k+1} = S_{k+1} - S_k > -l > -\infty \quad a.s. , \tag{2.38}$$

and

$$E(z_{k+1}/\mathcal{F}_k) \geq \epsilon \quad \text{a.s.} \tag{2.39}$$

where

$$z_k = y_k \mathbf{1}_{\{y_k < b\}} ,$$

then, for $S_0 > C$,

$$P(\tau = \infty) > 0 ;$$

remember that C appears in the definition of τ.

Proof : The line of argument is the same as that used to derive theorem 2.1.9 from theorem 2.1.8 and indeed is based on the exponential estimates obtained in theorem 2.1.7. Therefore, the details will be omitted. ■

2.2 Criteria for countable Markov chains

Let us consider a time homogeneous Markov chain $\mathcal{L}$ with a countable state space $\mathcal{A} = \{\alpha_i, i \geq 0\}$. The n-step transition probabilities will be denoted by $p^{(n)}_{\alpha_i\alpha_j}$ or, more briefly, by $p^{(n)}_{ij}$, with $p^{(1)}_{ij} \equiv p_{ij}$. $\mathcal{L}$ is supposed to be irreducible and aperiodic. The position of the chain at time n is ξ_n, as introduced in chapter 1.

Theorem 2.2.1 *The Markov chain $\mathcal{L}$ is recurrent if, and only if, there exist a positive function $f(\alpha), \alpha \in \mathcal{A}$, and a finite set A, such that*

$$E[f(\xi_{m+1}) - f(\xi_m)/\xi_m = \alpha_i] \leq 0 \ , \quad \forall \alpha_i \notin A \ , \tag{2.40}$$

and $f(\alpha_j) \to \infty$, when $j \to \infty$.

Proof: Let τ_i the $\mathcal{F}_n$-stopping time representing the epoch of first entry into the set A, given that $\xi_0 = \alpha_i \notin A$, i.e.

$$\tau_i = \inf\{n \geq 1 : \xi_n \in A/\xi_0 = \alpha_i\} .$$

We first prove the *if* assertion.

Let $S_n = f(\xi_n)$ and $\tilde{S}_n = f(\xi_{n\wedge\tau_i})$. Condition (2.40) entails that $\{\tilde{S}_n, \mathcal{F}_n\}$ is a positive supermartingale, since

$$E[\tilde{S}_{n+1}/\mathcal{F}_n] \leq \tilde{S}_n \quad \text{a.s.} \tag{2.41}$$

It is well known that there exists $\tilde{S}_\infty = \lim_{n\to\infty} \tilde{S}_n$, almost surely. Moreover, from Fatou's lemma,

$$E(\tilde{S}_\infty) \leq E(\tilde{S}_0) = S_0 = f(\xi_0)\,. \tag{2.42}$$

Suppose the chain is transient. As $f(\alpha_j) \to \infty$ when $j \to \infty$, there exist $m_0, \delta > 0$ and $\alpha_i \notin A$ such that, for any K, we have

$$P(\{f(\xi_m) > K\} \cap_{k=1}^{\infty} \{\xi_k \notin A\}/\xi_0 = \alpha_i) > \delta \ , \ \forall m > m_0\,. \tag{2.43}$$

But (2.43) yields

$$E(\tilde{S}_m/\xi_0 = \alpha_i) \to \infty \ , \ \text{as } m \to \infty\ , \tag{2.44}$$

which contradicts (2.42). So the chain is recurrent.

We shall now prove the *only if* assertion, i.e. the existence of a function $f(\alpha)$ satisfying (2.40), whenever $\mathcal{L}$ is recurrent. The states are now enumerated by the integers $0, 1, 2, \ldots$. Let $\tilde{\mathcal{L}}$ be the Markov chain with transition probabilities

$$\begin{cases} \tilde{p}_{00} &= 1, \\ \tilde{p}_{ij} &= p_{ij},\ \forall i \geq 1\,, \forall j \geq 0. \end{cases}$$

Let $\tilde{\xi}_m$ be the position of $\tilde{\mathcal{L}}$ at time m and denote by $\varphi_i(n)$ the probability that $\tilde{\mathcal{L}}$ ever reaches the set $\{n, n+1, \ldots\}$, given that $\tilde{\xi}_0 = i$.
Thus $\varphi_0(n) = 0$, $\forall n > 0$ and $\varphi_i(n) = 1$, $\forall i \geq n$. Moreover, since $\mathcal{L}$ is recurrent, we have

$$\lim_{n\to\infty} \varphi_i(n) = 0 \ , \ \forall i \geq 0. \tag{2.45}$$

Now we construct an increasing sequence of integers $\{n_i, i = 1, 2, 3, \ldots\}$ subject to the following conditions:

For any $k \geq 1$, we can choose n_k, such that $\varphi_i(n_k) < 2^{-k}, \forall i \leq k$.

This is possible owing to (2.45). Note, that for fixed n, $\varphi_i(n)$ is a function of i and, if we set $\varphi(\alpha_i) = \varphi_i(n)$, then (2.40) is satisfied for $A = \alpha_0$, i.e.

$$E[\varphi(\xi_{m+1}) - \varphi(\xi_m)/\xi_m = \alpha_j] \leq 0\ , \ \forall \alpha_j \neq \alpha_0.$$

Let us now define the function

$$f(\alpha_i) = \sum_{k=1}^{\infty} \varphi_i(n_k).$$

From the definition of $\varphi_i(n)$ and n_k, it follows that $\sum_{k=1}^{\infty} \varphi_i(n_k) < \infty$ and $f(\alpha_i)$ satisfies (2.40), as does $\varphi(\alpha_i) = \varphi_i(n_k)$. But for any fixed k, $\varphi_i(n_k) \to 1$ and $\sum_{k=1}^{\infty} \varphi_i(n_k) < \infty$, so that $\lim_{i\to\infty} f(\alpha_i) = +\infty$, by Fatou's lemma.

The proof of the theorem is concluded. ∎

Theorem 2.2.2 *The Markov chain $\mathcal{L}$ is transient, if and only if there exist a positive function $f(\alpha), \alpha \in \mathcal{A}$ and a set A such that the following inequalities are fulfilled:*

$$E[f(\xi_{m+1}) - f(\xi_m)/\xi_m = \alpha_i] \leq 0 \ , \ \forall \alpha_i \notin A \ , \tag{2.46}$$

$$f(\alpha_k) < \inf_{\alpha_j \in A} f(\alpha_j), \text{for at least one } \alpha_k \notin A \ . \tag{2.47}$$

Proof : The notation is the same as in theorem 2.40. Assume that (2.46) and (2.47) hold. Then (2.46) yields

$$E[f(\xi_{n\wedge\tau_k})] \leq E[f(\xi_0)] = f(\alpha_k) \ .$$

Suppose the chain not transient. Then $P(\tau_k < \infty) = 1$ and

$$\xi_{n\wedge\tau_k} \stackrel{a.s.}{\to} \xi_{\tau_k} \ .$$

Hence by using Fatou's lemma, we get

$$E[f(\xi_{\tau_k})] \leq E[f(\xi_0)] \leq f(\alpha_k) \ ,$$

which contradicts (2.47), since $\xi_{\tau_k} \in A$. Thus $P(\tau_k = \infty) > 0$ and the chain is transient.

To show that (2.46) and (2.47) are necessary, let us fix some arbitrary state α_0 and take $A = \{\alpha_0\}$. Define the function $f(\cdot)$ as follows:

$$\begin{cases} f(\alpha_0) & = & 1 \ , \\ f(\alpha_j) & = & P\{f(\xi_\nu) = \alpha_0, f(\xi_k) \neq \alpha_0, 1 \leq k < \nu/\xi_0 = \alpha_j\}, \ j \geq 1. \end{cases}$$

Then, clearly,

$$E[f(\xi_{m+1}) - f(\xi_m)/\xi_m = \alpha_j] = 0 \ , \ \forall \alpha_j \neq \alpha_0 \ .$$

Moreover, since the chain is assumed to be transient, $f(\alpha_j) < 1 = f(\alpha_0) \ , \ j \neq 0$.

The theorem is completely proved. ∎

Theorem 2.2.3 (Foster) *The Markov chain $\mathcal{L}$ is ergodic if and only if there exist a positive function $f(\alpha), \alpha \in \mathcal{A}$, a number $\epsilon > 0$ and a finite set $A \in \mathcal{A}$ such that*

$$E[f(\xi_{m+1}) - f(\xi_m)/\xi_m = \alpha_j] \leq -\epsilon \ , \ \alpha_j \notin A \ , \tag{2.48}$$

$$E[f(\xi_{m+1})/\xi_m = \alpha_i] < \infty \ , \ \alpha_i \in A \ . \tag{2.49}$$

Proof : First, let us prove the sufficiency. The notation is the same as in 2.2.1, but here the $\mathcal{F}_n$-stopping-times

$$\tau_i = \inf\{n \geq 1, \xi_n \in A/\xi_0 = \alpha_i\}$$

are defined for all $\alpha_i \in \mathcal{A}$. Let us fix $\alpha_i \notin A$ and define

$$S_n = f(\xi_n) \ , \ \tilde{S}_n = S_{n\wedge\tau_i} \ .$$

Rewriting (2.48) in the equivalent form

$$E[\tilde{S}_{m+1} - \tilde{S}_m/\xi_m = \alpha_j] \leq -\epsilon\mathbf{1}(\tau_j > m) \ , \ \alpha_j \notin A \ , \tag{2.50}$$

we can apply theorem 2.1.1 to get immediately

$$E(\tau_i) \ \leq \ \frac{f(\alpha_i)}{\epsilon} \ , \ \text{for all } \alpha_i \notin A \ . \tag{2.51}$$

Now, for any $\alpha_k \in A$, we have, using (2.51) and (2.49),

$$\begin{aligned} E(\tau_k) &= \sum_{\alpha_i \in A} p_{ki} + \sum_{\alpha_i \notin A} p_{ki} E[\tau_i + 1] \\ &= 1 + \sum_{\alpha_i \notin A} p_{ki} E(\tau_i) \leq 1 + \frac{1}{\epsilon} \sum_{\alpha_i \notin A} p_{ki} f(\alpha_i) < \infty \ . \end{aligned}$$

Thus we have shown that for all $\alpha_k \in A$, the mean return time to the finite set A is finite. This is equivalent to positive recurrence, since it implies that, for one (and thus for all) $\alpha_k \in A$, $m_{\alpha_k\alpha_k} < \infty$, according to the definition given in section 1.1. This shows the *if* part of the proposition.

To prove the necessity, we shall construct a function $f(\cdot)$ satisfying (2.48) and (2.49), assuming that $\mathcal{L}$ is ergodic.

Choose $A = \{\alpha_0\}$, where α_0 is a fixed state, and define

$$\begin{cases} f(\alpha_i) = E(\tau_i) \ , \ i > 0 \ , \\ f(\alpha_0) = 0 \ . \end{cases}$$

Then, it is straightforward to check that

$$E[f(\xi_{m+1}) - f(\xi_m)/\xi_m = \alpha_i] = -1 \ , \ \forall \alpha_i \neq \alpha_0 \ ,$$

which is another way of writing the system

$$E(\tau_i) = \sum_{j=1}^{\infty} p_{ij} E(\tau_j) + 1 \;, \quad i \neq 0 \,.$$

Moreover, since $m_{\alpha_0 \alpha_0} < \infty$, we also have

$$\sum_{j=0}^{\infty} p_{0j} E(\tau_j) < \infty \,,$$

which is nothing else but (2.49). The proof of the theorem is finished. ■

The next theorem is a generalization of Foster's theorem, just as theorem 2.1.2 was a generalization of theorem 2.1.1. It will be frequently used in the rest of the book.

Theorem 2.2.4 *The Markov chain $\mathcal{L}$ is ergodic if and only if there exist a positive function $f(\alpha), \alpha \in \mathcal{A}$, a number $\epsilon > 0$, a positive integer-valued function $k(\alpha), \alpha \in \mathcal{A}$, and a finite set A, such that the following inequalities hold:*

$$E[f(\xi_{m+k(\xi_m)}) - f(\xi_m)/\xi_m = \alpha_i] \leq -\epsilon k(\alpha_i), \; \alpha_i \notin A \,; \tag{2.52}$$

$$E[f(\xi_{m+k(\xi_m)})/\xi_m = \alpha_i] < \infty \;, \quad \alpha_i \in A \,. \tag{2.53}$$

Proof: Follows directly from theorem 2.1.2 and from the argument used in theorem 2.2.3. The details are omitted. ■

As an immediate consequence, we have

Corollary 2.2.5 *All the conditions of theorem 2.2.4, together with*

$$\sup_{\alpha \in \mathcal{A}} k(\alpha) = k < \infty \,,$$

are necessary and sufficient for $\mathcal{L}$ to be ergodic. ■

Theorem 2.2.6 *For an irreducible Markov chain $\mathcal{L}$ to be non-ergodic, it is sufficient that there exist a function $f(\alpha), \alpha \in \mathcal{A}$, and constants C and d such that*

(i) $E[f(\xi_{m+1}) - f(\xi_m)/\xi_m = \alpha] \geq 0$, *for every m, all $\alpha \in \{f(\alpha) > C\}$, where the sets $\{\alpha : f(\alpha) > C\}$ and $\{\alpha : f(\alpha) \leq C\}$ are non-empty;*

(ii) $E[| f(\xi_{m+1}) - f(\xi_m) | /\xi_m = \alpha] \leq d$, *for every m, $\forall \alpha \in \mathcal{A}$.*

Proof: The random sequence $S_n = f(\xi_n)$, $\xi_0 =$ constant, $f(\xi_0) > C$, satisfies the conditions of theorem 2.1.3, with $\tau = \inf\{n \geq 1 : f(\xi_n) \leq C\}$. Thus $E(\tau) = \infty$ and $\mathcal{L}$ is non-ergodic. ■

Theorem 2.2.7 *For an irreducible Markov chain $\mathcal{L}$ to be transient, it suffices that there exist a positive function $f(\alpha), \alpha \in \mathcal{A}$, a bounded integer-valued positive function $k(\alpha)$, $\alpha \in \mathcal{A}$, and numbers $\epsilon, C > 0$, such that, setting $A_c = \{\alpha : f(\alpha) > C\} \neq \emptyset$, the following conditions hold:*

(i) $\sup_{\alpha \in \mathcal{A}} k(\alpha) = k < \infty$;

(ii) $E(f(\xi_{m+k(\xi_m)}) - f(\xi_m)/\xi_m = \alpha_i) \geq \epsilon$, $\forall\, m$, *for all* $\alpha_i \in A_c$;

(iii) *for some* $d > 0$, *the inequality* $\mid f(\alpha_i) - f(\alpha_j) \mid > d$ *implies* $p_{ij} = 0$.

Proof : Still denote by τ the time of first entry of the sequence $S_n = f(\xi_n)$ into [0, C]. From condition (ii) above, it is not difficult to see by induction that there exists ξ_0, such that $f(\xi_0) = C + dk$. We introduce the random sequence $N_0 = k(\xi_0), N_{i+1} = N_i + k(\xi_i)$. Then, condition (ii) can be rewritten as

$$E(S_{N_i}/S_{N_{i-1}} > C) \geq S_{N_{i-1}} + \epsilon\,,$$

and we are entitled to apply theorem 2.1.9, which yields $P(\tau = \infty) > 0$. Thus $\mathcal{L}$ is transient and the theorem is proved. ■

Theorem 2.2.8 *For an irreducible Markov chain $\mathcal{L}$ to be null recurrent, it suffices that there exist two functions $f(x)$ and $\varphi(x), x \in X$, and a finite subset $A \in X$, such that the following conditions hold:*

(i) $f(x) \geq 0$, $\varphi(x) \geq 0$, $\forall x \in X$.

(ii) *For some positive* α, γ, *with* $1 < \alpha \leq 2$,

$$f(x) \leq \gamma[\varphi(x)]^\alpha, \ \forall x \in X\,.$$

(iii) $\lim_{x_i \to \infty} \varphi(x_i) = \infty$ *and*

$$\sup_{x \notin A} f(x) > \sup_{x \in A} f(x)\,.$$

(iv) (a) $E[f(\xi_{n+1}) - f(\xi_n)/\xi_n = x] \geq 0$, $\forall x \notin A$;

(b) $E[\varphi(\xi_{n+1}) - \varphi(\xi_n)/\xi_n = x] \leq 0$, $\forall x \notin A$;

(c) $\sup_{x \in X} E[\mid \varphi(\xi_{n+1}) - \varphi(\xi_n) \mid^\alpha / \xi_n = x] = C < \infty$.

Proof : Let us suppose the existence of $f(\cdot)$ and $\varphi(\cdot)$. Conditions (i), (iii) and (iv)(b) on $\varphi(x)$ show immediately, by using theorem 2.2.1, that $\mathcal{L}$ is recurrent. We shall now assume that $\mathcal{L}$ is ergodic, in order to get a contradiction, thus proving the null recurrence. Let us denote

$$\begin{cases} a_n &= \varphi(\xi_n),\ b_n = f(\xi_n)\ , \\ \tau &= \inf\{n > 0 : \xi_n \in A/\xi_0 \notin A\}\ , \\ \tilde{a}_n &= a_{n\wedge\tau}\ ,\ \tilde{b}_n = b_{n\wedge\tau}\ . \end{cases}$$

Since $\mathcal{L}$ was assumed to be ergodic, $E(\tau) < \infty$. It will be convenient to choose ξ_0 to be a constant and $\xi_0 \notin A$. The set A being finite, we have

$$\sup_{x\in A} \varphi(x) < \infty\ ,\ \sup_{x\in A} f(x) < \infty\ .$$

Since $P(\tau < \infty) = 1$, there exist two random variables $\tilde{a}$ and $\tilde{b}$, such that

$$\tilde{a}_n = \varphi(\xi_{n\wedge\tau}) \overset{a.s.}{\longrightarrow} \tilde{a}\ ,\ 0 \le \tilde{a} \le \sup_{x\in A} \varphi(x)\ ,$$

$$\tilde{b}_n = f(\xi_{n\wedge\tau}) \overset{a.s.}{\longrightarrow} \tilde{b}\ ,\ 0 \le \tilde{b} \le \sup_{x\in A} f(x)\ .$$

Moreover, theorem 2.1.5 shows that the random variables $\tilde{a}_n^\alpha$ are uniformly integrable and converge to $\tilde{a}^\alpha$ in the L^1-sense. Using now condition (ii) in the statement of theorem 2.2.8, we get

$$\tilde{b}_n = f(\xi_{n\wedge\tau}) \le \gamma\tilde{a}_n^\alpha\ .$$

Thus the family $\{\tilde{b}_n, n \ge 0\}$, dominated by a uniformly integrable family, is also uniformly integrable. This shows that $\tilde{b}$ is the L^1-limit of $\tilde{b}_n$ and

$$\lim_{n\to\infty} E(\tilde{b}_n) = E(\tilde{b})\ \le\ \sup_{x\in A} f(x)\ . \tag{2.54}$$

On the other hand, condition (iv)(a) shows that $\tilde{b}_n$ is a submartingale and

$$E[\tilde{b}_n/\xi_0 = \alpha_i] \ge f(\xi_0) = f(\alpha_i)\ ,\ \forall\alpha_i \notin A\ ,\ \forall n \ge 0\ . \tag{2.55}$$

Condition (iii) allows us to choose i in (2.55), such that

$$f(\alpha_i) > \sup_{x\in A} f(x)\ .$$

Doing so, we get from the estimate (2.54), which does not depend on the initial position ξ_0,

$$\lim_{n\to\infty} E(\tilde{b}_n/\xi_0 = \alpha_i) \le \sup_{x\in A} f(x)\ ,$$

and this last inequality contradicts (2.55). Thus necessarily $E(\tau) = \infty$ and the proof of theorem 2.2.8 is completed. ■

3

Explicit construction of Lyapounov functions

The simplest idea to come to mind is that, in order to use the criteria of section 2.2, one must exhibit explicit Lyapounov functions. Fortunately enough, this can be done for a large number of cases. In this chapter, we give typical examples where Lyapounov functions can be found and verified with elementary (but sometimes very tedious) calculations.
In section 3.1, the simplest case is considered, in which, however, arises the notion of the induced chain, fundamental for chapter 4.
In section 3.2, the main definition of a space homogeneous random walk in $\mathbf{Z}_+^N$ is given. The classification for $N = 2$ is obtained in sections 3.3 and 3.4.
A special but famous type of random walk in $\mathbf{Z}_+^N$ is given by Jackson networks in section 3.5, for which necessary and sufficient conditions of ergodicity are proved, by constructing explicit Lyapounov functions.
In section 3.6, we examine important one-dimensional examples, when the drifts are asymptotically zero. The last section 3.7 presents some results pertaining to the invariance principle.

3.1 Markov chains in a half-strip

Here we consider an MC $\mathcal{L}$, defined on the state space $\mathbf{Z}_+ \times \mathbf{Z}_n$, $\mathbf{Z}_n = \{1, \ldots, n\}$. The states are denoted by $\alpha = (x, i), x \in \mathbf{Z}_+, i = 1, \ldots, n$.

Assumption $\mathbf{A}_0$ (*Homogeneity*) For almost all (i.e. except for a finite number of) points $\alpha = (x, i)$, the transition probabilities from (x, i) to $(x + k, j)$ do not depend on x and thus can be denoted by p_k^{ij}.

Assumption $\mathbf{A}_1$ (*Lower boundedness*) $p_k^{ij} = 0$, for $k < d$, for some $d > -\infty$.

We now introduce, albeit in the simplest situation, the new notion of *induced chain*, intensively used in the next chapter.

Definition 3.1.1 *The* induced chain $\mathcal{L}_{ind}$ *for the random walk* $\mathcal{L}$ *is a finite MC, with state space* $\mathbf{Z}_n$ *and governed by the transition probabilities*

$$q_{ij} = \sum_k p_k^{ij} . \tag{3.1}$$

Let $A_1, \ldots, A_m$ be the essential classes for $\mathcal{L}_{ind}$ (which itself may be not irreducible). Let $\mathcal{L}_s$ be the induced MC on A_s and let $\pi_i,\ i \in A_s$, be the stationary distribution of $\mathcal{L}_s$, $s = 1, \ldots, m$.
Let $M(i) = \sum_{j,k} k\, p_k^{ij}$ be the mean jump from the point $\alpha = (x, i)$, for almost all x.
Define also the mean drift in the x-direction, for $s = 1, \ldots, m$,

$$M_s = \sum_{i,j,k} k\pi_i p_k^{ij} = \sum_i \pi_i\, M(i) \ \text{ where } i,j \text{ run over } A_s \, .$$

We also introduce the following:
Assumption $\mathbf{A_2}$ (*Boundedness of moments*) $M(i) < \infty$ for all i.

Theorem 3.1.2 *The following classification holds:*

(i) $\mathcal{L}$ *is ergodic if, and only if,* $M_s < 0,\ \forall s$;
(ii) $\mathcal{L}$ *is recurrent if, and only if,* $M_s \leq 0,\ \forall s$.

Proof We consider only the case when all states of the induced chain $\mathcal{L}_{ind}$ are essential, i.e. $m = 1$, and then write $M = M_1$, leaving obvious generalizations to the reader.

(i) First, assume $M = 0$. We shall prove that this case corresponds to null recurrence, by finding a function

$$f(\alpha) = f((x, i)) = x + a_i,$$

which satisfies the equalities

$$\sum_\beta p_{\alpha\beta} f(\beta) - f(\alpha) = 0, \text{ for almost all } \alpha \, . \tag{3.2}$$

For x sufficiently large and $\alpha = (x, i)$, equation (3.2) is thus equivalent to

$$\sum_{jk} p_k^{ij}(k + a_j) - a_i = M(i) + \sum_j q_{ij}\, a_j - a_i = 0 \ , \ i = 1, \ldots, n \, . \tag{3.3}$$

This system of equations has a solution if, and only if, the vector $(M(1),\dots,M(n))$ is orthogonal to any solution of the adjoint homogeneous system, i.e. when

$$M = \sum_{i=1}^{n} \pi_i \, M(i) = 0. \tag{3.4}$$

Therefore, the function $f(\alpha)$ satisfies the conditions of theorem 2.2.1, so that the MC $\mathcal{L}$ is recurrent. On the other hand, $f(\alpha)$ satisfies the conditions of theorem 2.2.6: thus $\mathcal{L}$ is non-ergodic and, consequently, is null recurrent.

(ii) If $M < 0$, then we introduce the function $f(\alpha) = f((x,i)) = x + a_i$, satisfying the inequalities

$$\sum_{\beta} p_{\alpha\beta} f(\beta) - f(\alpha) < -\epsilon, \quad \text{for some } \epsilon > 0 \text{ and almost all } \alpha. \tag{3.5}$$

Now we apply Foster's criterion of theorem 2.2.3, hence proving the ergodicity. Note that here the lower boundedness is not needed.

(iii) If $M > 0$, then we use a function satisfying the inequalities

$$\sum_{\beta} p_{\alpha\beta} f(\beta) - f(\alpha) > \epsilon, \quad \text{for some } \epsilon > 0 \text{ and almost all } \alpha, \tag{3.6}$$

and the transience follows from theorems 2.2.7 and 2.1.10.
Theorem 3.1.2 is proved. ■

3.1.1 Generalizations and problems

Let $\mathcal{A}$ be a denumerable set and consider a Markov chain defined on the countable state space $\mathcal{A} \times \mathbf{Z}$ and transition probabilities $p((\alpha,n) \to (\beta,m))$, homogeneous in the second component, i.e. depending only on α , β and $m-n$. We assume also that there exists a constant $d < \infty$, such that $p((\alpha,n) \to (\beta,m)) = 0$, for $| m-n | > d$. Let us put

$$M(\alpha) = \sum_{(\beta,m)} (m-n) p((\alpha,n) \to (\beta,m)) \, .$$

We define the induced chain as in (3.1), the quantity M as in (3.4), and we assume that the induced chain is irreducible, aperiodic and ergodic. A set $A = \mathcal{A} \times (\mathbf{Z} - \mathbf{Z}_+) \subset \mathcal{A} \times \mathbf{Z}$ is called

(i) *positive recurrent*, if the mean time of reaching A from any state $(\alpha, n) \in \mathcal{A} \times \mathbf{Z}$, $n > 0$, is finite;

(ii) *recurrent* if the probability of reaching A, from any state $(\alpha, n) \in \mathcal{A} \times \mathbf{Z}$, $n > 0$, is equal to 1;

(iii) *transient* if the above probability is less than 1.

We have obtained above the complete classification, when $\mathcal{A}$ is finite and the induced chain is irreducible, namely

(a) A is positive recurrent if, and only if, $M < 0$;

(b) A is recurrent if, and only if, $M \leq 0$.

Below, we prove (b) in a more general situation. The reader can show easily that A is positive recurrent if $M < 0$. We can formulate the following

Problem : When is it true that A is not positive recurrent if $M = 0$?

Let ξ_n be an irreducible aperiodic ergodic Markov chain, with countable state space $\mathcal{A}$, and $g(\alpha)$ be a bounded integer-valued function on $\mathcal{A}$. Let $\tau_1, \tau_2, \ldots$ be the random times of successive visits to some fixed state i. We define

$$x_T = c + \sum_{t=0}^{T-1} g(\xi_t), \; c > 0 \, .$$

Theorem 3.1.3 *Under the above conditions, the set A is recurrent for the chain (ξ_n, x_{n+1}) if, and only if, $M \leq 0$.*

Proof Setting $\eta_j = x_{\tau_{j+1}} - x_{\tau_j}$, it is sufficient to prove that $E(\eta_j) = 0$, when $M = 0$. To that end, we shall use some results and notation borrowed from [Chu67], sections I.13 and I.14, where ${}_i p_{ij}^{(k)}$ is the taboo probability, starting from the state i, of entering the state j at the k-th step, without hitting i in between. Then the η_j's are i.i.d. random variables and $E[\| \eta_j \|]$ is finite, for all $j \geq 1$. Setting ${}_i p_{ij}^* = \pi_j / \pi_i$, we have

$$E(\eta_i) = \sum_{k,j} g(j) \, {}_i p_{ij}^{(k)} = \sum_j g(j) \, {}_i p_{ij}^* = \sum_j g(j) \frac{\pi_j}{\pi_i} = 0.$$

Thus x_{τ_i} enters 0 a.s. and so does x_n and theorem 3.1.3 is proved. ■

Remark The so-called *periodic* random walk in [Key84] pertains to the theory of this section. A random walk in $\mathbf{Z}$ or $\mathbf{Z}^+$ is called *periodic*, with period U, if the transition probabilities satisfy

$$p_{\alpha\beta} = p_{\alpha+U, \beta+U} \, .$$

(On $\mathbf{Z}^+$, this property needs to be verified only for α sufficiently large.) This periodic-random-walk problem can immediately be reduced to that of the strip $\mathbf{Z} \times \{1, \ldots, U\}$ or of $\mathbf{Z}^+ \times \{1, \ldots, U\}$. Note that a result similar to that of [Key84] was already contained in [Mal72b].

3.2 Random walks in $\mathbf{Z}_+^N$: main definitions and interpretation

Let us consider a discrete time homogeneous Markov chain $\mathcal{L}$ which is assumed to be irreducible and aperiodic unless otherwise stated. We introduce the notation which will be ubiquitous in the sequel.
The set of states is $\mathbf{Z}_+^N = \{(z_1, \ldots, z_N) : z_i \geq 0 \text{ integers}\}$, $p_{\alpha\beta}^k$ are the k-step transition probabilities of $\mathcal{L}$, with $p_{\alpha\beta}^1 = p_{\alpha\beta}$. Also, let

$$M^k(\alpha) = (M_1^k(\alpha), \ldots, M_N^k(\alpha))$$

be the vector of mean jumps from the point α in k steps. We shall write

$$M(\alpha) \stackrel{\text{def}}{=} M^1(\alpha) = \sum_\beta (\beta - \alpha) p_{\alpha\beta}.$$

For any $1 \leq i_1 < \ldots < i_k \leq N, k \geq 1$, a *face* of

$$\mathbf{R}_+^N = \{(r_1, \ldots, r_N) : r_i \geq 0 \text{ real}\}$$

is, by definition,

$$B^\wedge \equiv \wedge(i_1, \ldots, i_k) =$$
$$\{(r_1, \ldots, r_N) : r_i > 0, i \in \{i_1, \ldots, i_k\}; r_i = 0, i \notin \{i_1, \ldots, i_k\}\}.$$

In spite of a slight ambiguity in the notation, we shall sometimes write $i \in B^\wedge$ or $i \in \wedge$, when $i \in \{i_1, \ldots, i_k\}$. It is important to emphasize that a face does not include its boundary. Quite naturally, $\wedge_1 \subset \wedge$ will mean exactly $B^{\wedge_1} \subset \overline{B^\wedge}$
We shall frequently consider random walks satisfying the two following conditions already encountered in section 3.1:

Condition A_0 (*Maximal homogeneity*) For any $\wedge$ and for all $a \in B^\wedge \cap \mathbf{Z}_+^N$,

$$p_{\alpha\beta} = p_{\alpha+a, \beta+a}, \quad \forall \alpha \in B^\wedge \cap \mathbf{Z}_+^N, \ \forall \beta \in \mathbf{Z}_+^N.$$

Thus we can write $p_{\alpha\beta} = p(\wedge; \beta - \alpha)$.

Condition A_1 (*Boundedness of the jumps*)

$$p_{\alpha\beta} = 0 \text{ for } \parallel \alpha - \beta \parallel > d\,,$$

where d is a strictly positive constant and

$$\parallel \alpha \parallel = \max_i \mid \alpha_i \mid, \alpha = (\alpha_1, \ldots, \alpha_N).$$

Condition A_0 ensures that $p_{\alpha\beta} = 0$, if $\beta_i - \alpha_i < -1$, for at least one i. We define now the *first vector field* on $\mathbf{R}_+^N$ to be constant on any $\wedge$ and equal to

$$M_\wedge \equiv M(\alpha),\ \alpha \in \wedge\,.$$

Unless otherwise stated, we assume that conditions A_1 and A_0 are in force.

As an example, consider a class of networks with N queues. All customers in the same queue are identical (no types, no marks), so that the state of the network at time t is completely defined by the vector

$$\alpha = (n_1(t), \ldots, n_N(t)),$$

where n_i is the number of customers in queue i. One can have various transitions (including sychronization constraints)

$$\alpha = (n_1(t), \ldots, n_N(t)) \to \beta = (n_1(t) + i_1, \ldots, n_N(t) + i_N),$$

with respective intensities $\lambda_{\alpha\beta}$, yielding different values of the vector

$$\beta - \alpha = (i_1, \ldots, i_N), \tag{3.7}$$

for nonzero $\lambda_{\alpha\beta}$'s. For example, in the famous continuous time Jackson network, not more than two components of the vector (3.7) can be different from zero. But in the case of *Fork-Join* systems, involving bulk-arrivals and bulk-services (see e.g [BM89, ICB88]), the vector (3.7) may have up to N nonzero components. As usual, the random walks corresponding to these systems are defined, for w sufficiently small, by

$$p_{\alpha\beta} = w\lambda_{\alpha\beta},\ \alpha \neq \beta,\ p_{\alpha\alpha} = 1 - \sum_{\beta\neq\alpha} p_{\alpha\beta}\,. \tag{3.8}$$

This class of networks is not sufficient to get arbitrary random walks satisfying A_0 and A_1, because the transition intensities on the faces are simply restrictions of the transition intensities inside $\mathbf{Z}_+^N$ (a kind of *meta-continuity*). In fact the most general random walk, in the class we have introduced in this section, can be depicted by a queueing network with interactions between the nodes, where *interaction* means that $\lambda_{\alpha\beta}$ depend also on which nodes of the network are empty, if any. More exactly, $\lambda_{\alpha\beta} \equiv \lambda_{\alpha\beta}(\wedge; \beta - \alpha)$, which means that $\lambda_{\alpha\beta}$ is a function of $\beta - \alpha$ and of the face $\wedge$ to which α belongs. Thus networks with interactions provide in fact all Markov chains defined on $\mathbf{Z}_+^N$, subject to conditions A_0 and A_1. Examples of such networks are given not by Jackson networks, but for example by *Buffered ALOHA, coupled-processors* [Szp90, FI79], ..., etc. A new class of networks for data bases will also be considered in section 5.9.

3.3 Classification of random walks in $\mathbf{Z}_+^2$

Consider a discrete time homogeneous irreducible and aperiodic MC $\mathcal{L} = \{\xi_\eta, n \geq 0\}$. Its state space is the lattice in the positive quarter-plane $\mathcal{Z}_+^2 = \{(i,j) : i,j \geq 0, \text{integers}\}$ and it satisfies the recursive equation

$$\xi_{n+1} = [\xi_n + \theta_{n+1}]^+ ,$$

where the distribution of θ_{n+1} depends only on the position of ξ_n in the following way (*maximal space homogeneity*):

$$p\{\theta_{n+1} = (i,j)/\xi_n = (k,l)\} = \begin{cases} p_{ij}, & \text{for } k,l \geq 1 , \\ p'_{ij}, & \text{for } k \geq 1, l = 0 , \\ p''_{ij}, & \text{for } k = 0,\ l \geq 1 , \\ p^0_{ij}, & \text{for } k = l = 0 . \end{cases}$$

Moreover we shall make, for the one-step transition probabilities, the following assumptions:

Condition A *(Lower boundedness)*

$$\begin{cases} p_{ij} = 0, & \text{if} \quad i < -1 \quad \text{or } j < -1 ; \\ p'_{ij} = 0, & \text{if} \quad i < -1 \quad \text{or } j < 0 ; \\ p''_{ij} = 0, & \text{if} \quad i < 0 \quad \text{or } j < -1 . \end{cases}$$

Condition B *(First moment condition)*

$$E[\| \theta_{n+1} \| /\xi_n = (k,l)] \leq C < \infty, \quad \forall (k,l) \in \mathbf{Z}_+^2 ,$$

where $\| z \|, z \in \mathbf{Z}_+^2$, denotes the euclidean norm and C is an arbitrary but strictly positive number.

Notation We shall use lower case greek letters $\alpha, \beta, \ldots$ to denote arbitrary points of $\mathbf{Z}_+^2$, and then $p_{\alpha\beta}$ will mean the one-step transition probabilities of the Markov chain $\mathcal{L}$ and $\alpha > 0$ means

$$\alpha_x > 0,\ \alpha_y > 0,\ \text{for } \alpha = (\alpha_x, \alpha_y) .$$

Also, from the homogeneity conditions, one can write

$$\theta_{n+1} = (\theta_x, \theta_y), \ \text{ given that } \xi_\eta = (x,y).$$

Define the vector

$$M(\alpha) = (M_x(\alpha), M_y(\alpha))$$

of the one-step mean jumps (drifts) from the point α. Setting

$$\alpha = (\alpha_x, \alpha_y),\ \beta = (\beta_x, \beta_y) ,$$

we have

$$M_x(\alpha) = \sum_\beta p_{\alpha\beta}(\beta_x - \alpha_x) ,$$

$$M_y(\alpha) = \sum_\beta p_{\alpha\beta}(\beta_y - \alpha_y) .$$

Condition **B** ensures the existence of $M(\alpha)$, for all $\alpha \in \mathbf{Z}_+^2$. By the homogeneity condition **A**, only four drift vectors are different from zero:

$$M(\alpha) = \begin{cases} M, & \text{for } \alpha_x, \alpha_y > 0 , \\ M', & \text{for } \alpha = (\alpha_x, 0),\ \alpha_x > 0 ; \\ M'', & \text{for } \alpha = (0, \alpha_y),\ \alpha_y > 0 ; \\ M_0, & \text{for } \alpha = (0,0) . \end{cases}$$

Remark:

(i) All our results remain valid if a finite number of transition probabilities are arbitrarily modified.

(ii) Given $\xi_n = \alpha$, the components of θ_{n+1} might be taken bounded from below not by -1, but by some arbitrary number $-K > -\infty$, provided that

First, we keep the maximal homogeneity for the drift vectors $M(\alpha)$ introduced above (i.e. four of them only different);
Secondly, the second moments and the covariance of the one-step jumps inside $\mathbf{Z}_+^2$, i.e. from any point $\alpha > 0$, are kept constant. These last facts will emerge more clearly in the course of the study.

Theorem 3.3.1 *Assume conditions* **A** *and* **B** *are satisfied.*

(a) *If* $M_x < 0$, $M_y < 0$, *then the Markov chain* $\mathcal{L}$ *is*

(i) *ergodic if*

$$\begin{cases} M_x M_y' - M_y M_x' & < 0, \\ M_y M_x'' - M_x M_y'' & < 0; \end{cases}$$

(ii) *non-ergodic if either*

$$M_x M_y' - M_y M_x' \geq 0 \quad \textit{or} \quad M_y M_x'' - M_x M_y'' \geq 0.$$

(b) *If* $M_x \geq 0, M_y < 0$, *then the Markov chain* $\mathcal{L}$ *is*

(i) *ergodic if*

$$M_x M_y' - M_y M_x' < 0 ;$$

(ii) *transient if*

$$M_x M_y' - M_y M_x' > 0 .$$

(c) *(Case symmetric to case (b)) If* $M_y \geq 0, M_x < 0$*, then the Markov chain* $\mathcal{L}$ *is*

(i) *ergodic if*

$$M_y M_x'' - M_x M_y'' < 0 ;$$

(ii) *transient if*

$$M_y M_x'' - M_x M_y'' > 0 .$$

(d) *If* $M_x \geq 0,\ M_y \geq 0,\ M_x + M_y > 0$*, then the Markov chain is transient.*

Assuming also that the jumps are bounded with probability 1, some stronger results can be derived.

Theorem 3.3.2 *Let there exist* $d > 0$*, such that* $\| \theta_n \| \leq d$ *a.s., for all* n*, and assume conditions A and B hold.*

(a) *If* $M_x < 0, M_y < 0$*, then the Markov chain* $\mathcal{L}$ *is*

(i) *transient if either*

$$M_x M_y' - M_y M_x' > 0 \quad \text{or} \quad M_y M_x'' - M_x M_y'' > 0 ;$$

(ii) *null recurrent if either*

$$\begin{cases} M_x M_y' - M_y M_x' = 0, \\ M_y M_x'' - M_x M_y'' \leq 0, \end{cases}$$

or

$$\begin{cases} M_x M_y' - M_y M_x' \leq 0, \\ M_y M_x'' - M_x M_y'' = 0 . \end{cases}$$

(b) *If* $M_x \geq 0, M_y < 0$*, then the Markov chain* $\mathcal{L}$ *is null recurrent if*

$$M_x M_y' - M_y M_x' = 0 .$$

(c) *If* $M_x < 0, M_y \geq 0$*, then the Markov chain* $\mathcal{L}$ *is null recurrent if*

$$M_y M_x'' - M_x M_y'' = 0 .$$

Proof of theorem 3.3.1 : Introduce the following real functions on $\mathbf{Z}_+^2$:

$$\begin{cases} Q(x,y) &= ux^2 + vy^2 + wxy, \\ f(x,y) &= Q^{1/2}(x,y), \\ \Delta f(x,y) &= Q^{1/2}(x+\theta_x, y+\theta_y) - Q^{1/2}(x,y), \end{cases}$$

where $(x,y) \in \mathbf{Z}_+^2$ and u, v, w are unspecified constants, to be properly chosen later, but subject to the constraints $u, v > 0$, $4uv > w^2$, so that the quadratic form Q is positive definite. First, we shall prove the ergodicity in the case (a(i)).

Lemma 3.3.3

$$E[\Delta f(x,y)] = \frac{x[2uE(\theta_x) + wE(\theta_y)] + y[wE(\theta_x) + 2vE(\theta_y)]}{2f(x,y)} + o(1) , \tag{3.9}$$

where $o(1) \to 0$ *as* $(x^2+y^2) \to \infty$.

Proof

$$E[\Delta f(x,y)] = E[f(x+\theta_x, y+\theta_y) - f(x,y)] \tag{3.10}$$

$$= f(x,y)E\Big[\Big(1 + \frac{x(2u\theta_x + w\theta_y) + y(w\theta_x + 2v\theta_y) + Q(\theta_x,\theta_y)}{Q(x,y)}\Big)^{1/2} - 1\Big].$$

Let us write $E(\Delta f(x,y))$ in the form

$$E[\Delta f(x,y)] = \psi_1(x,y) + \psi_2(x,y),$$

where

$$\begin{aligned} \psi_1(x,y) &= E[\Delta f(x,y)\mathbf{1}_{\{|\theta_x+\theta_y|\le z\}}] , \\ \psi_2(x,y) &= E[\Delta f(x,y)\mathbf{1}_{\{|\theta_x+\theta_y|> z\}}] , \end{aligned}$$

z being some positive real number.
Take (x,y) such that $x^2+y^2 = \mathcal{D}^2$ and $z = \epsilon_1\mathcal{D}$, for ϵ_1 sufficiently small and $\mathcal{D}$ large. Then, for $\mid \theta_x + \theta_y \mid < z$ and sufficiently small ϵ_1, we have

$$\Big|\frac{x(2u\theta_x + w\theta_y) + y(w\theta_x + 2v\theta_y) + Q(\theta_x,\theta_y)}{Q(x,y)}\Big| < 1 .$$

Upon applying now the simple inequality

$$(1+t)^\beta \le 1 + \beta t, \quad \text{for } \mid t \mid < 1 \text{ and } 0 < \beta \le 1,$$

and taking $\epsilon_1 < D^{-\alpha}$ with $\alpha > 1/2$, it follows that

$$\psi_1(x,y) \quad =$$
$$f(x,y)E\Big[\frac{x(2u\theta_x+w\theta_y)+y(w\theta_x+2v\theta_y)+Q(\theta_x,\theta_y)}{2Q(x,y)}\mathbf{1}_{\{|\theta_x+\theta_y|\le z\}}\Big]+o(1)$$
$$= \frac{E[(x(2u\theta_x+w\theta_y)+y(w\theta_x+2v\theta_y))\mathbf{1}_{\{|\theta_x+\theta_y|\le z\}}]}{2f(x,y)}$$
$$+E\Big[\frac{Q(\theta_x,\theta_y)}{2f(x,y)}\mathbf{1}_{\{|\theta_x+\theta_y|\le z\}}\Big]+o(1).$$

Noting that, if $E[|\ \xi\ |] = C < \infty$, for any arbitrary random variable ξ, then

$$E[\xi\ \mathbf{1}_{\{|\xi|>z\}}] = o(1) \text{ and } E[\xi^2\ \mathbf{1}_{\{|\xi|\le z\}}] \le Cz\ ,$$

we get

$$\psi_1(x,y) = \frac{xE(2u\theta_x+w\theta_y)+yE(w\theta_x+2v\theta_y)}{2f(x,y)} + \frac{o(1)(x+y)}{f(x,y)} + \epsilon_1 O(1)\ , \tag{3.11}$$

where $O(1)$ represents, as usual, a bounded quantity.
Let us now estimate $\psi_2(x,y)$. For fixed u, v, w, ϵ_1, there exists a constant a such that

$$\Delta f(x,y)\mathbf{1}_{\{|\theta_x+\theta_y|>z\}} \le a|\theta_x+\theta_y|\mathbf{1}_{\{|\theta_x+\theta_y|>z\}}\ .$$

Hence

$$\psi_2(x,y) \le aE[|\theta_x+\theta_y|\mathbf{1}_{\{\theta_x+\theta_y|>z\}}] = o(1), \text{ as } z\to\infty\ . \tag{3.12}$$

The lemma now follows from the estimates (3.11) and (3.12). ■

Let us continue the proof of case (a(i)). Lemma 3.3.3 shows that, if there exist $u, v > 0$ and $w^2 < 4uv$, such that, for some $\epsilon_2 > 0$ and all $(x,y) \in \mathbf{Z}_+^2 - A$, where A is a finite set,

$$\begin{cases} 2uE(\theta_x)+wE(\theta_y) < -\epsilon_2, \\ wE(\theta_x)+2vE(\theta_y) < -\epsilon_2, \end{cases} \tag{3.13}$$

then, for some $\mathcal{D}, \epsilon > 0$ and all (x,y) with $x^2+y^2 > \mathcal{D}^2$, we have

$$E(\Delta f(x,y)) < -\epsilon\ . \tag{3.14}$$

Therefore, when (3.13) holds, the random walk is ergodic, by using theorem 2.2.3 (Foster's criterion). Let us rewrite inequalities (3.13) in terms of the drifts on the axes and in the internal part of $\mathbf{Z}_+^2$.

$$\begin{cases} 2u\ M_x + wM_y < -\epsilon_2, \\ 2v\ M_y + wM_x < -\epsilon_2, \\ 2u\ M_x' + wM_y' < -\epsilon_2, \\ 2v\ M_y'' + wM_x'' < -\epsilon_2\ . \end{cases} \tag{3.15}$$

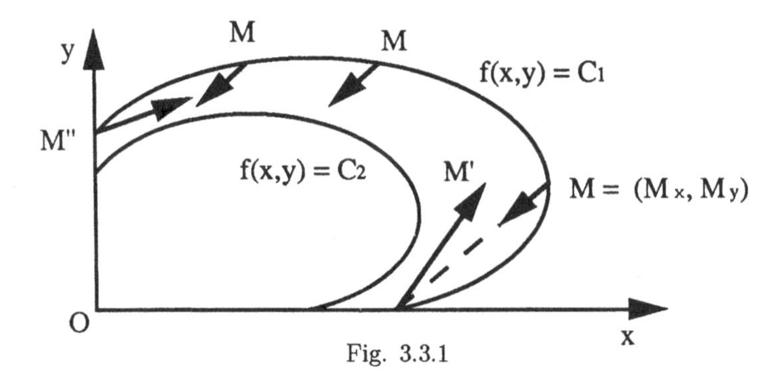

Fig. 3.3.1

It is easy to show that, if

$$\begin{cases} M_x < 0, \\ M_y < 0, \\ M_x M'_y - M_y M'_x < 0, \\ M_y M''_x - M_x M''_y < 0, \end{cases} \tag{3.16}$$

then there exist $u, v > 0$ and $w^2 < 4uv$, such that (3.15) is satisfied for some $\epsilon_2 > 0$, thus proving case $a(i)$. We can give a geometrical illustration of this result on Fig. 3.3.1, where inequalities (3.16) hold, so that the chain is ergodic.

The cases (b(i)) and (c(i)) are analogous to (a(i)). Indeed, if

$$\begin{cases} M_x \geq 0\,, \\ M_y < 0\,, \\ M_x M'_y - M_y M'_x < 0\,, \end{cases} \quad \text{or} \quad \begin{cases} M_x < 0\,, \\ M_y \geq 0\,, \\ M_y M''_x - M_x M''_y < 0\,, \end{cases}$$

we show that there exist $u, v > 0$ and $w^2 < 4uv$, such that (3.15) holds, so that the chain is ergodic in both cases. Fig. 3.3.2, shows the situation corresponding to (b(i)).

Now we shall prove the non-ergodicity in (a(ii)). Assume that

$$\begin{cases} M_x < 0, \\ M_y < 0, \\ M_x M'_y - M_y M'_x \geq 0\,. \end{cases} \tag{3.17}$$

As shown in Fig 3.3.3, there exists a linear function $f(x,y) = ax + by$, such that, for all $\alpha = (x,y)$ with $ax + by \geq C$, we have

$$f(\alpha + M(\alpha)) \geq f(\alpha) + \epsilon, \quad \text{for some } C, \epsilon \geq 0\,,$$

and the non-ergodicity immediately follows from theorem 2.2.6.

The proof of the transience in (b(ii)), (c(ii)) and (d) is more difficult.

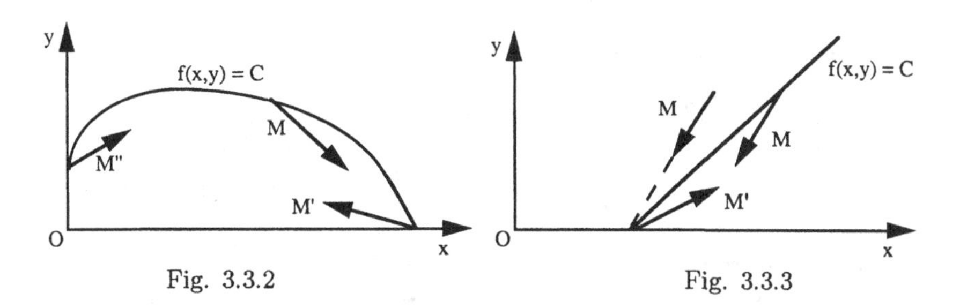

Fig. 3.3.2 Fig. 3.3.3

In a preliminary step, we shall discuss the principle of *local linearity* for random walks with bounded jumps.

Principle of local linearity Let the state space $\mathcal{A}$ of a Markov chain $\mathcal{L}$ be a countable subset of $\mathbf{R}^N$ and the function f_α , $\alpha \in \mathcal{A}$, be the restriction of some real function $f(x)$ defined on $\mathbf{R}^N$. Assume also $f_\alpha \geq C > -\infty$ and introduce the sets

$$\mathcal{D}_\epsilon^- = \{x : f(x) < \epsilon, x \in \mathbf{R}^N\} \text{ and } \mathcal{D}_\epsilon^+ = \{x : f(x) > \epsilon, x \in \mathbf{R}^N\} .$$

Suppose the jumps of $\mathcal{L}$ are bounded, i.e., for any $\alpha \in A$, there exists a number d_α such that

$$\| \beta - \alpha \| > d_\alpha \text{ implies that } p_{\alpha\beta} = 0, \ \forall \beta \in \mathcal{A} .$$

If the function $f(x)$ is linear, then it is easy to see that $\alpha + M(\alpha) \in \mathcal{D}^-_{f(\alpha)-\epsilon}$ if, and only if,

$$\sum_{\beta \in \mathcal{A}} p_{\alpha\beta}(f_\beta - f_\alpha) < -\epsilon . \tag{3.18}$$

Analogously, $\alpha + M(\alpha) \in \mathcal{D}^+_{f(\alpha)+\epsilon}$ if, and only if,

$$\sum_{\beta \in \mathcal{A}} p_{\alpha\beta}(f_\beta - f_\alpha) > \epsilon . \tag{3.19}$$

Let now $f(x)$ be arbitrary.

Lemma 3.3.4 *If* $\alpha + M(\alpha) \in \mathcal{D}^-_{f(\alpha)-5\epsilon}$ *and the condition*

$$\inf_{\varphi} \sup_{\substack{\tilde{\alpha} \in \mathbf{R}^n \\ \|\tilde{\alpha} - \alpha\| \leq d_\alpha}} | f(\tilde{\alpha}) - \varphi(\tilde{\alpha}) | < \epsilon \tag{3.20}$$

holds, where inf *is taken over all linear functions* φ, *then inequality (3.18) is valid.*

Proof Let $\varphi(x)$ be a linear function such that

$$\sup_{\substack{\tilde{\alpha}\in\mathbf{R}^n \\ \|\tilde{\alpha}-\alpha\|\leq d_\alpha}} \mid f(\tilde{\alpha}) - \varphi(\tilde{\alpha}) \mid < \epsilon . \tag{3.21}$$

Then we have the decomposition

$$\sum_\beta p_{\alpha\beta}(f(\beta) - f(\alpha))$$
$$= \sum_\beta p_{\alpha\beta}(\varphi_\beta - \varphi_\alpha) + \sum_\beta p_{\alpha\beta}(f_\beta - \varphi_\beta) + \sum_\beta p_{\alpha\beta}(\varphi_\alpha - f_\alpha) . \tag{3.22}$$

Using (3.21), we can write

$$\Big|\sum_\beta p_{\alpha\beta}(f_\beta - \varphi_\beta)\Big| \leq \epsilon \text{ and } \Big|\sum_\beta p_{\alpha\beta}(\varphi_\alpha - f_\alpha)\Big| \leq \epsilon .$$

Hence

$$\sum_\beta p_{\alpha\beta}(\varphi_\beta - \varphi_\alpha) = \varphi(\alpha + M(\alpha)) - \varphi(\alpha)$$
$$\leq [\varphi(\alpha + M(\alpha)) - f(\alpha + M(\alpha))]$$
$$+[f(\alpha) - \varphi(\alpha)] + [f(\alpha + M(\alpha)) - f(\alpha)]$$
$$\leq \epsilon + \epsilon - 5\epsilon = -3\epsilon .$$

Thus, we obtain from (3.22)

$$\sum_{\beta\in A} p_{\alpha\beta}(f(\beta) - f(\alpha)) \leq -\epsilon .$$

The proof of lemma 3.3.4 is concluded. ■

Lemma 3.3.5 *If* $\alpha + M(\alpha) \in \mathcal{D}^+_{f(\alpha)+5\epsilon}$ *and the condition*

$$\inf_\varphi \sup_{\substack{\tilde{\alpha}\in\mathbf{R}^n \\ \|\tilde{\alpha}-\alpha\|\leq d_\alpha}} \mid f(\tilde{\alpha}) - \varphi(\tilde{\alpha}) \mid < \epsilon$$

holds, then inequality (3.19) is valid.

The proof of this result mimics completely that of lemma 3.3.4. ■

It is natural to call the statement of lemmas 3.3.4 and 3.3.5 *the principle of local linearity.* Instead of verifying conditions 3.18 and 3.19, this principle allows us automatically to use smooth level curves, for a Lyapounov function transversal to the one-step mean jump vector field. In many concrete situations, $\varphi(\tilde{\alpha})$ will be the tangent to the level curve at the point α, in particular when the level curve behaves as $\|\alpha\|^p, p \leq 1$.

Proof of case (d) Let us consider first the case of bounded jumps. We begin with a geometrical construction of the function $f(x, y), (x, y) \in \mathbf{R}^2_+$.

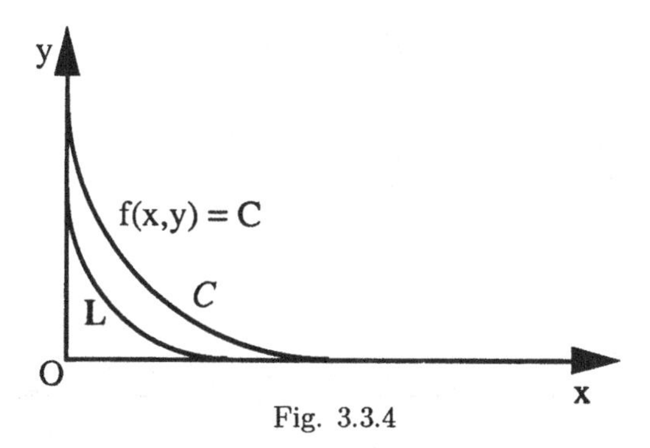

Fig. 3.3.4

We draw the quarter-circle L of radius 1, tangent to the axes x and y, as shown in figure 3.3.4. For all points (x, y) belonging to the quarter-circle L, we put $f(x, y) = 1$. Then, by scaling, we define the function $f(x, y)$ in the whole quarter-plane $\mathbf{R}_+^2$. More exactly, we put, for $(x, y) \in L$ and all $r \geq 0$,

$$f(rx, ry) = r\,.$$

It is clear that the level curve C corresponding to $f(x, y) = c$ is the circle of radius c with centre at the point (c, c). So we can apply the principle of local linearity to the function f. Hence, for any $d, \epsilon > 0$, there exists $\mathcal{D} > 0$ such that, for any $\alpha \in \mathbf{R}_+^2, \| \alpha \| > \mathcal{D}$,

$$\inf_{\varphi} \sup_{\substack{\tilde{\alpha} \in \mathbf{R}^n \\ \|\tilde{\alpha} - \alpha\| \leq d}} | f(\tilde{\alpha}) - \varphi(\tilde{\alpha}) | < \epsilon,$$

where inf is taken over all linear functions φ.

Moreover, it appears that all mean jump vectors (except maybe on the axes, in the particular situation when M' or M'' points towards the origin along its respective axis) in the direction of the increasing values of $f(\alpha)$. Hence, taking $\mathcal{D}$ such that, for $\| \alpha \| > \mathcal{D}$,

$$\alpha + M(\alpha) \in \mathcal{D}^+_{f(\alpha)+5\epsilon}\,,$$

we obtain, from lemma 3.3.5,

$$\sum_{\beta \in \mathcal{A}} p_{\alpha\beta}(f(\beta) - f(\alpha)) > +\epsilon \quad , \text{ for } \| \alpha \| > \mathcal{D}\,, \alpha \notin Ox \cup Oy\,,$$

and the transience in the case (d), with bounded jumps, follows from theorem 2.2.7.

Let us deal now with the case in which jumps are bounded from below,

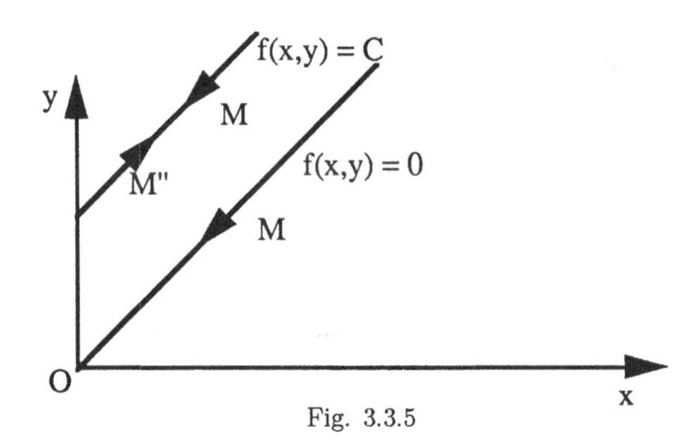

Fig. 3.3.5

but not from above. We take the same function $f(\alpha) \equiv f(x,y)$ as above and consider the two sequences of random variables

$$Y_k = f(\xi_{k+1}) - f(\xi_k) \text{ and } Z_k = Y_k \mathbf{1}_{\{Y_k < B\}} .$$

It is easy to verify that, for B sufficiently large, these sequences satisfy conditions of theorem 2.1.10, for $f(\xi_0)$ sufficiently large. The case (d) is thus completely proved.

Cases (b(ii)) and (c(ii)) could be proved in an entirely similar way and so are left as exercises. The proof of theorem 3.3.1 is complete. ■

Proof of Theorem 3.3.2 : The transience in the case (a(i)) can be proved as in the case (a(ii)) of theorem 3.3.1 but using theorem 2.2.7.
(1) *Consider first the case* (a(ii)). The proof of the non-ergodicity is then rather simple: it is indeed a direct consequence of theorem 2.2.6, by using the linear function shown in figure 3.3.5. Let us now prove the recurrence.
Introduce the following real functions, defined on $\mathbf{Z}_+^2$:

$$Q(x,y) = ux^2 + vy^2 + wxy \text{ and } f(x,y) = \log[Q(x,y) + 1] ,$$

where u, v, w are unspecified constants (to be properly chosen later) satisfying $u, v > 0$ and $4uv \geq w^2$, so that $Q(x,y)$ shall be a positive semi-definite quadratic form. We have

$$\begin{aligned} E[\Delta f] &= E[f(x + \theta_x,\ y + \theta_y)] - f(x,y) \\ &= E\big[\log[1 + \frac{x(2u\theta_x + w\theta_y) + y(w\theta_x + 2v\theta_y) + Q(\theta_x, \theta_y)}{1 + Q(x,y)}]\big] \\ &= \frac{x[2uE(\theta_x) + wE(\theta_y)] + y[wE(\theta_x) + 2vE(\theta_y)] + E[Q(\theta_x, \theta_y)]}{1 + Q(x,y)} \end{aligned}$$

$$-\frac{E\big[[x(2u\theta_x+w\theta_y)+y(w\theta_x+2v\theta_y)+Q(\theta_x,\theta_y)]^2\big]}{2[1+Q(x,y)]^2}$$

$$+o[\frac{1}{x^2+y^2}] \,. \tag{3.23}$$

Let us first assume that

$$\begin{cases} M_xM_y'-M_yM_x'=0, \\ M_yM_x''-M_xM_y''<0\,. \end{cases} \tag{3.24}$$

Then we choose the constants u, v, w, to satisfy the system

$$\begin{cases} 2uM_x'+wM_y'=0, \\ 2uM_x+wM_y=0, \\ 2vM_y+wM_x<0, \\ 2vM_y''+wM_x''<0. \end{cases} \tag{3.25}$$

It turns out that (3.25) is equivalent to the simpler system

$$\begin{cases} u = -\dfrac{wM_y'}{2M_x'} = -\dfrac{wM_y}{2M_x}, \\ -\dfrac{2vM_y}{M_x} < w < -\dfrac{2vM_y''}{M_x''}\,. \end{cases} \tag{3.26}$$

Put

$$v>0\,,\ \frac{w}{v}=-\frac{2M_y}{M_x}+\epsilon\,,\ \frac{u}{v}=\left[\frac{M_y}{M_x}\right]^2-\frac{\epsilon M_y}{2M_x}\,. \tag{3.27}$$

Choosing $\epsilon>0$ sufficiently small, it follows from (3.27) that (3.25) and (3.26) are satisfied, together with $Q(x,y)>0,\ \forall(x,y)$. We shall now estimate $E(\Delta f)$.

- Assume first $y>l$, where l is sufficiently large. From the boundedness of the jumps it follows that

$$\begin{cases} E[Q(\theta_x,\theta_y)] < A\,, \text{for some constant } A\,, \\ x[2uE(\theta_x)+wE(\theta_y)] = 0\,, \\ y[wE(\theta_x)+2vE(\theta_y)] < -\epsilon_2 l\,,\ \text{for some } \epsilon_2>0. \end{cases}$$

Hence, for l sufficiently large, $E(\Delta f(x,y))<0$.

- Let now $y\le l$. Then we have

$$\begin{aligned} E(\Delta f) &\le \frac{E[Q(\theta_x,\theta_y)]}{1+Q(x,y)}-\frac{x^2[4u^2E(\theta_x^2)+4wuE(\theta_x\theta_y)+w^2E(\theta_y^2)]}{2[1+Q(x,y)]^2} \\ &\qquad +o(\frac{1}{x^2+y^2}) \\ &= \frac{1}{ux^2}[-uE(\theta_x^2)-wE(\theta_x\theta_y)-(\frac{w^2}{2u}-v)E(\theta_y^2)] \\ &\qquad +o(\frac{1}{x^2}) \end{aligned}$$

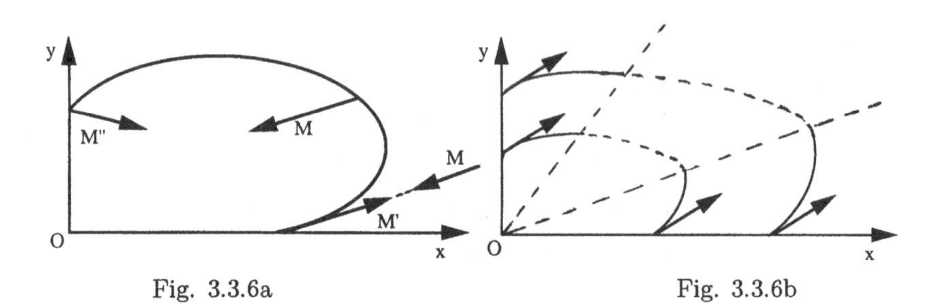

Fig. 3.3.6a Fig. 3.3.6b

$$= \frac{1}{ux^2}\left[-E[Q(\theta_x,\theta_y)] - \frac{1}{2u}(w^2-4uv)E(\theta_y^2)\right] + o(\frac{1}{x^2}) \,. \quad (3.28)$$

Choosing ϵ small in (3.27) and inserting the corresponding v, w, u into (3.28), we see that $w^2 - 4uv \to 0$, if $\epsilon \to 0$, and

$$E[Q(\theta_x,\theta_y)] \geq E[(\sqrt{u}\theta_x - \sqrt{v}\theta_y)^2] > B > 0 \,,$$

for some B not depending on ϵ. Hence, for sufficiently large x, $E(\Delta f) \leq 0$. Finally, when (3.24) holds, we have been able to construct a function f fulfilling the conditions of theorem 2.2.1, so that the Markov chain is recurrent.

It is useful to illustrate the above proof geometrically. The equipotential lines of the function f are the ellipses $f(x,y) = C$, shown in figure 3.3.6(a). They are tangent to the vectors M' at the points of the x-axis and the vectors M and M'' point towards their interior.

The case

$$\begin{cases} M_xM'_y - M_yM'_x = 0 \,, \\ M_yM''_x - M_xM''_y = 0 \,, \end{cases} \quad (3.29)$$

could be handled along the same lines, although with additional technical difficulties. In fact, one can show that the function $f(x,y)$ increases as $\log(x^2+y^2)$, but the level curves would consist of two ellipses, connected by a smooth arc, as depicted in figure 3.3.6(b). To describe the obstacles in trying to produce a direct approach, as above, we shall prove the recurrence in a particular case, which requires an additional (strictly speaking not necessary) assumption, involving a relationship between second-order quantities.

Choose u, v, w satisfying

$$v > 0 \,, \quad \frac{w}{v} = -\frac{2M_y}{M_x} \,, \quad \frac{u}{v} = \left[\frac{M_y}{M_x}\right]^2 .$$

For these values of u, v, w, which are compatible with (3.29), we set

$$\begin{aligned} Q(x,y) &= T^2(x,y), \text{ where } T(x,y) = \sqrt{u}\,x - \sqrt{v}\,y\,, \\ L(x,y) &= ax + by, \text{ where } a \text{ and b are positive constants }, \\ R(x,y) &= Q(x,y) + L(x,y) \text{ and } f(x,y) = \log[R(x,y)]\,. \end{aligned}$$

From now on, we take $\|x+y\| > D$, for $D > 0$ large enough, and it is important to note that Q is only positive semi-definite. For any function $g(x,y)$, we introduce the convenient notation

$$\Delta g(x,y) = \Delta g \stackrel{\text{def}}{\equiv} g(x+\theta_x, y+\theta_y) - g(x,y)\,.$$

Thus we have

$$E[\Delta T(x,y)] = 0 \text{ and } E[\Delta L(x,y)] = aE[\theta_x] + bE[\theta_y]\,.$$

Hence

$$E[\Delta R(x,y)] = E[(\Delta T(x,y))^2 + \Delta L(x,y)]$$

and

$$\begin{aligned} &E[\Delta f(x,y)] \\ &= \frac{2(T^2+L)E[(\Delta T)^2 + \Delta L] - E[(2T\Delta T + (\Delta T)^2 + \Delta L)^2]}{2R^2(x,y)} \\ &\qquad + o(E\Big[\Big(\frac{\Delta R}{R}\Big)^2\Big]) \\ &= \frac{-2T^2E[(\Delta T)^2 - \Delta L] + 2LE[(\Delta T)^2 + \Delta L] - 4TE[\Delta T(\Delta L + (\Delta T)^2)]}{2R^2(x,y)} \\ &\qquad - \frac{\delta^2}{2R^2(x,y)} + o(E\Big[\Big(\frac{\Delta R}{R}\Big)^2\Big])\,. \end{aligned}$$

(We have omitted the argument (x,y) in most of the functions.) It turns out that the right-hand side of the above equation (in which δ^2 denotes a positive quantity uniformly bounded with respect to (x,y)) can be rendered negative outside a compact set, i.e. for D large enough, if we assume the condition

$$\inf\left\{\frac{E[(\Delta T')^2]}{M'_x}, \frac{E[(\Delta T'')^2]}{M''_x}\right\} > \frac{E[(\Delta T^2)]}{M_x}\,,$$

(which is parasitic in the sense that it is not needed in the statement of the theorem) with an obvious notation ($'$ refers to the Ox-axis, $''$ to the Oy-axis and no *prime* symbol means the interior of the quarter-plane). Geometrically, the level curves are here *parabolas*. The case (a(ii)) is finished.

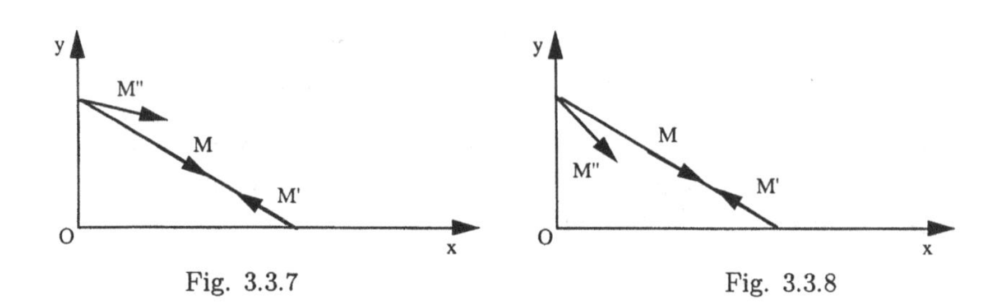

Fig. 3.3.7 Fig. 3.3.8

(2) *case b.* There are two subcases.

- First, the situation shown in figure 3.3.7. Then the recurrence can be proved as in case (a(ii)) and the details are left to the reader. On the other hand, taking

$$\varphi(\alpha) \equiv \varphi(x,y) = px + qy, \text{ for } \alpha = (x,y) \text{ , with } p > 0,\ q > 0 \text{ ,}$$

we can choose p and q so that

$$\begin{cases} \varphi(M) &= pM_x + qM_y = 0 \text{ ,} \\ \varphi(M') &= pM'_x + qM'_y = 0 \text{ ,} \\ \varphi(M'') &= pM''_x + qM''_y > 0 \text{ .} \end{cases} \tag{3.30}$$

Thus, $\varphi(\xi_n)$ is a positive submartingale. Hence

$$E[\varphi(\xi_{n+1}) \mid \xi_n = (x,y)] \geq \varphi(x,y)$$

and theorem 2.2.6 tells us that the chain is non-ergodic and, consequently, is null recurrent.

- Consider now the case of figure 3.3.8. This is a more difficult situation, which we shall solve by using theorem 2.2.8, together with the principle of local linearity. Here, it is possible to choose $p, q > 0$, such that $\varphi(\xi_n)$ becomes a positive supermartingale and the Markov chain $\mathcal{L}$ is recurrent by theorem 2.2.1. Our goal now is to prove the non-ergodicity.

Define the regions

$$\begin{aligned} \mathcal{B} &= \Big\{(x,y) \in \mathbf{R}^2_+ \cap \{y < ax\}\Big\} \text{ , for some } a \text{ with } 0 < a < \infty \text{ ,} \\ \mathcal{B}^c &= \Big\{(x,y) \in \mathbf{R}^2_+ \cap \{y \geq ax\}\Big\} \text{ .} \end{aligned}$$

For $(x,y) \in \mathcal{B}$, we introduce the functional of quadratic form,

$$f(x,y) = (ux^2 + vy^2 + xy)^\delta \text{ ,} \tag{3.31}$$

where $u \geq 0,\ v \geq 0,\ 0 < \delta < 1$, and recall the notation

$$\begin{aligned} Q(x,y) &= ux^2 + vy^2 + xy\,, \\ \Delta Q(x,y) &= Q(x+\theta_x, y+\theta_y) - Q(x,y) \\ &= u\theta_x^2 + v\theta_y^2 + \theta_x\theta_y + (y+2ux)\theta_x + (x+2vy)\theta_y\,. \end{aligned} \tag{3.32}$$

Our purpose is to estimate the quantity

$$\begin{aligned} H(x,y) &\stackrel{\text{def}}{\equiv} E[f(\xi_{n+1}) - f(\xi_n)/\xi_n = (x,y)] \\ &= E[(Q(x,y) + \Delta Q(x,y))^\delta - Q^\delta(x,y)]\,. \end{aligned} \tag{3.33}$$

Lemma 3.3.6 *There exist δ, $0 < \delta < 1$ and a constant D such that*

$$H(x,y) = \delta Q^{\delta-1}(x,y)[E(\Delta Q(x,y)) + (\delta-1)O(1) + o(1)] \tag{3.34}$$

for all (x,y), such that $(x^2+y^2) > D^2$, where, as usual, $|O(z)| < K|z|$ and $o(1)$ tends to zero when $D \to \infty$.

Proof : By Taylor's formula, we have

$$\begin{aligned} H(x,y) &= \delta E[\Delta Q(x,y)(Q(x,y) + \gamma(x,y)\Delta Q(x,y))^{\delta-1}] \\ &= \delta Q^{\delta-1}(x,y)[E[\Delta Q(x,y)] + \psi(x,y)], \end{aligned} \tag{3.35}$$

where $\gamma(x,y)$ is a random variable such that $0 < \gamma(x,y) < 1$.
Note that $Q + \gamma\Delta Q \geq 0$ and

$$\psi(x,y) = E\left[\Delta Q(x,y)\left[\left(1 + \frac{\gamma(x,y)\Delta Q(x,y)}{Q(x,y)}\right)^{\delta-1} - 1\right]\right]\,.$$

It suffices to prove $\psi(x,y) = (\delta-1)O(1) + o(1)$. In fact, the result of the lemma is immediate from the definition (3.33), after using $\|(1+z)^{\delta-1} - 1\| = (\delta-1)O(z)$, for z sufficiently small, since, from the boundedness of the jumps,

$$\frac{(\Delta Q(x,y))^2}{Q(x,y)} < A < \infty\,,\ \forall (x,y) \neq (0,0)\,.$$

The lemma is proved. ■

Let us continue the proof of the theorem. From (3.33), it follows that

$$\begin{aligned} &E[\Delta Q(x,y)] \\ &= xE[2u\theta_x + \theta_y] + yE[2v\theta_y + \theta_x] + E[Q(\theta_x,\theta_y)] \\ &= x(2uE(\theta_x) + E(\theta_y)) + y(2vE(\theta_y) + E(\theta_x)) \\ &= \quad uE(\theta_x^2) + vE(\theta_y)^2 + E(\theta_x\theta_y)\,. \end{aligned} \tag{3.36}$$

Now we shall find $\mathcal{D}, u, v > 0$, such that

$$E[\Delta Q(x,y)] > 0 \ , \ \text{for } (x,y) \in \mathcal{B} \text{ and } (x^2+y^2) \geq \mathcal{D}^2 \ .$$

Choose

$$\begin{cases} u &= -\dfrac{M_y}{2M_x} = -\dfrac{M'_y}{2M'_x} \ , \\ v &= -\dfrac{M_x}{2M_y} - \epsilon \ , \end{cases} \tag{3.37}$$

with ϵ sufficiently small. Then we distinguish two regions:

• First, the set $\{(x,y) \ : x > 0, \ y > 0\}$. Then (3.36) yields

$$\begin{aligned} E[\Delta Q(x,y)] &= x[-\frac{M_y}{M_x}M_x + M_y] + y[-M_x - 2\epsilon M_y + M_x] \\ &\quad + u\lambda_x + v\lambda_y + R \\ &= -2\epsilon y M_y + u\lambda_x + v\lambda_y + R \ , \end{aligned} \tag{3.38}$$

where

$$\lambda_x = E(\theta_x)^2, \ \lambda_y = E(\theta_y)^2, \ R = E(\theta_x\theta_y) \ .$$

• Secondly, the set $\{(x,y) \ : x > 0 \ , y = 0\}$. We obtain now

$$E[\Delta Q(x,y)] = u\lambda'_x + v\lambda'_y + R', \tag{3.39}$$

where $\lambda'_x = E(\theta_x)^2, \ \lambda'_y = E(\theta_y)^2, R' = E(\theta_x\theta_y)$.

Setting $z = -\dfrac{M_y}{M_x} = \tan\varphi > 0$, we get

$$\begin{aligned} u\lambda'_x + v\lambda'_y + R' &= \frac{(\tan\varphi)E(\theta_x)^2}{2} + \frac{(\cot\varphi)E(\theta_y)^2}{2} + E(\theta_x\theta_y) - \epsilon\lambda'_y \\ &= \frac{1}{2z}[z^2E(\theta_x)^2 + 2z \ E(\theta_x\theta_y) + E(\theta_y)^2] - \epsilon\lambda'_y \ . \end{aligned}$$

It follows from Schwartz's inequality $(E(\theta_x\theta_y))^2 \leq E(\theta_x^2) \ E(\theta_y^2)$, that

$$z^2E(\theta_x^2) + 2zE(\theta_x\theta_y) + E(\theta_y^2) > 0 \ . \tag{3.40}$$

Thus, using (3.38), (3.39), (3.40) and the assumption $M_y < 0 \ , \ M_x \geq 0$, we can choose $\epsilon > 0$, such that

$$E[\Delta Q(x,y)] > 0, \quad \text{for } x > 0, \ y \geq 0 \ .$$

Moreover, $E[\Delta Q(x,y)]$ grows as a linear function of y.

In figure 3.3.9, we show the level curve $Q(x,y) = c$. For $(x,y) \in \mathcal{B}$, this curve is an arc of an ellipse. For $(x,y) \in \mathcal{B}^c$, we draw the circle tangent

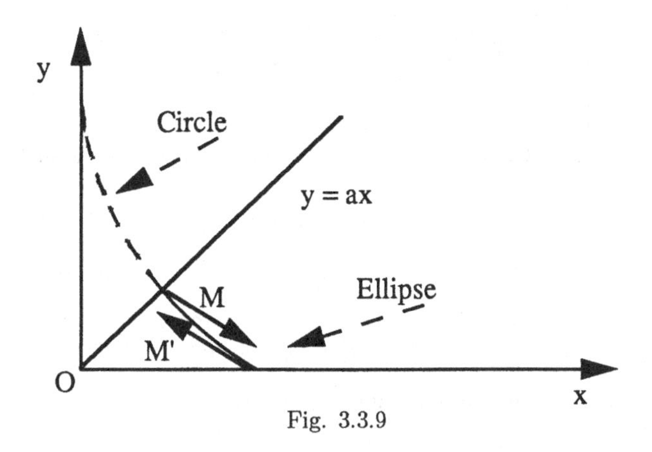

Fig. 3.3.9

to the vertical axis $x = 0$, which is a smooth continuation of the curve $Q(x,y) = c$. The union of these two curves will be denoted by L.

We have shown that for (x,y) , $x > 0$, $y > 0$, the quantity $E[\Delta Q(x,y)]$ grows linearly in y. So, for sufficiently large c, the vector M, drawn from the intersection point of L and of the straight line $y = ax$, is oriented in the direction of the increasing values of $Q(x,y)$ (see figure 3.3.9) and, therefore, also looks on the same side of L as the points belonging to the region $\mathcal{B}^c$.

Let us construct now the function $\tilde{f}$ defined as follows:

$$\begin{cases} \tilde{f}(x,y) & = & c \ , & \text{for } (x,y) \in L \ , \\ \tilde{f}(rx,ry) & = & r^2 c \ , & \text{for } (x,y) \in L \ , r \geq 1 \ . \end{cases} \tag{3.41}$$

We will prove the existence of δ, $0 < \delta < 1$ and $\mathcal{D}$, such that

$$E[\tilde{f}^\delta(x+\theta_x, y+\theta_y)] - \tilde{f}^\delta(x,y) \geq 0, \text{ for } (x,y) \in \mathbf{Z}_+^2 : x^2+y^2 \geq \mathcal{D}^2. \tag{3.42}$$

First, we will show that, for sufficiently large $\mathcal{D}$ and $(x,y) : x^2+y^2 \geq \mathcal{D}^2$,

$$E[\tilde{f}(x+\theta_x, y+\theta_y)] - \tilde{f}(x,y) \geq 0 \ . \tag{3.43}$$

For $(x,y) \in \mathcal{B}$, $\tilde{f}(x,y) = Q(x,y)$ and (3.43) has already been proved.

For $(x,y) \in \mathcal{B}^c$, inequality (3.43) follows easily from the principle of local linearity, which yields

$$E[\tilde{f}^{1/2}(x+\theta_x, y+\theta_y)] - \tilde{f}^{1/2}(x,y) \geq 0 \ , \ \text{for } (x,y) \in \mathcal{B}^c \ . \tag{3.44}$$

Now using (3.44) and Jensen's inequality, we can write

$$E[\tilde{f}^{\delta}(\ x+\theta_x+y+\theta_y)]-\tilde{f}^{\delta}(x,y)\geq 0\ ,$$
$$\text{for all } \delta,\ \frac{1}{2}<\delta\leq 1\ ,(x,y)\in \mathcal{B}^c,\ \text{and } x^2+y^2\geq \mathcal{D}^2, \quad (3.45)$$

where $\mathcal{D}$ is chosen sufficiently large.
To prove (3.42), if suffices now to apply lemma 3.3.6. In fact (3.34) shows that, taking δ close to 1, the sign of $H(x,y)$ is the same as that of $E[\Delta Q(x,y)]$. But we have already shown that

$$E[\Delta Q(x,y)]>0\ ,\ \text{for } (x,y)\in \mathcal{B}\ \text{ and } x^2+y^2>\mathcal{D}^2,\ \text{for some } \mathcal{D}\ .$$

We have finally proved that there exist $\mathcal{D}>0$ and $\frac{1}{2}<\delta<1$, such that (3.42) holds, for all $(x,y)\in \mathbf{Z}_+^2\cap\{x^2+y^2\geq \mathcal{D}^2\}$.

Now we are in a position to apply theorem 2.2.8, remembering that we are dealing with the case of figure 3.3.8. Then there exist $p,q>0$, such that the function $\varphi(\xi_n)$, where $\varphi(x,y)=px+qy$ is a positive supermartingale, i.e. condition (iv)(b) of theorem 2.2.8 holds. Since the jumps are bounded, condition (iv)(c) also holds. The following argument shows that condition (ii) also holds. Indeed, for any fixed $u,v,p,q>0$, there exists $d>0$, such that

$$ux^2+vy^2+xy<d(px+qy)^2\ ,$$

whence,

$$\tilde{f}(x,y)<d_1(\varphi(x,y))^{\delta}\ ,\ \text{for some } d_1>0\ .$$

Since we have shown above that conditions (i) and (iii) were also fulfilled, null recurrence follows from theorem 2.2.8.
This concludes the proof of case (b) and of theorem 3.3.2. ■

3.4 Zero drifts

We consider the Markov chain $\mathcal{L}$ introduced in the previous section, but satisfying the stronger

Condition B *(Second moment condition)*

$$E[\|\ \theta_{n+1}\ \|^2\ /\xi_n=(k,l)]\leq B<\infty\ ,\forall(k,l)\in \mathbf{Z}_+^2\ .$$

Until recently nothing was precisely known for the case $M=0$. In fact, this problem, in many respects, is of a very different nature. In particular, intuition does not provide us with any evidence that the random walk could be ergodic, when $M=0$.
There exist at least four methods which would allow us to solve some particular cases of the problem:

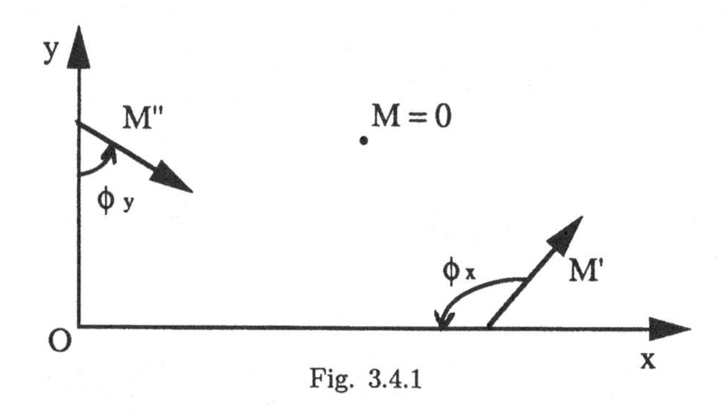

Fig. 3.4.1

(1) *The analytic approach*, as in [FI79, Mal70, Nau89]. Now there are only some preliminary results into this direction.
(2) *The semi-analytic approach*, by using well known explicit results about one-dimensional random walks. For instance, when the random walk inside the quarter-plane is the *composition* of two independent random walks along both axes, then it is easy to show that the mean time to reach the boundary is infinite, in which case, for any values of the parameters p'_{ij}, p''_{ij}, the chain is not ergodic if $M = 0$.
(3) *The method of Lyapounov functions.* This seems to be the most general approach and we will use it here.
(4) *The approach via diffusion processes.* There was a lot of work done on the same problems for diffusion processes in $\mathbf{R}_+^2$ (see [RW88, VW85, Wil85]). Their intimate connections with the discrete case will be discussed in a forthcoming monograph.

There is a crucial difference between the cases $M \neq 0$ and $M = 0$: indeed, the case $M \neq 0$ is in a sense locally *linear* and $M = 0$ is locally *quadratic.* The local second-order effects are well caught by functionals of quadratic Lyapounov functions.

For $M = 0$, we will obtain the ergodicity conditions in terms of the second moments and the covariance of the one-step jumps inside $\mathbf{Z}_+^2$,

$$\lambda_x = \sum_{ij} i^2 p_{ij} \ , \ \lambda_y = \sum_{ij} j^2 p_{ij} \ , R = \sum_{ij} ij p_{ij} \ ,$$

and of the angles, anticlockwise oriented, ϕ_x, ϕ_y shown on figure 3.4.1. Here ϕ_x is the angle between M' and the negative x-axis, ϕ_y is the angle between M'' and the negative y-axis. Thus, if $\phi_x \neq \pi/2$ and $\phi_y \neq \pi/2$, then

$$\tan \phi_x = -\frac{M_y'}{M_x'} \; , \tan \phi_y = -\frac{M_x''}{M_y''} \; .$$

Theorem 3.4.1

(i) *If* $\phi_x \geq \frac{\pi}{2}$ *or* $\phi_y \geq \frac{\pi}{2}$ *, then the random walk* $\mathcal{L}$ *is not ergodic.*

(ii) *If* $\phi_x < \frac{\pi}{2}$ *and* $\phi_y < \frac{\pi}{2}$*, then the random walk* $\mathcal{L}$ *is*

- *ergodic if*

$$\lambda_x \tan \phi_x + \lambda_y \tan \phi_y + 2R \equiv -\lambda_x \frac{M_y'}{M_x'} - \lambda_y \frac{M_x''}{M_y''} + 2R < 0; \quad (3.46)$$

- *non-ergodic if*

$$\lambda_x \tan \phi_x + \lambda_y \tan \phi_y + 2R > 0 \; . \quad (3.47)$$

(iii) *If (3.47) holds together with* $\phi_x + \phi_y \leq \frac{\pi}{2}$*, then the random walk is null recurrent.*

Remark 1 It follows easily from the statement of the theorem that the mean first entrance time of $\mathcal{L}$ into the boundary, when starting from some arbitrary point $\alpha > 0$ at finite distance, is finite (resp. infinite) if $R < 0$ (resp. $R > 0$), since in this case the vectors M' and M'' can be properly chosen to satisfy (3.46) (resp. (3.47)).

Remark 2 It is clear from the formulation of the theorem that we do not consider the limiting situation

$$\lambda_x \tan \phi_x + \lambda_y \tan \phi_y + 2R = 0 \; ,$$

which would impose further assumptions of third order.

Proof We introduce the linear function $\varphi : \; R_+^2 \rightarrow R_+$, such that, for any $\gamma = (x, y)$,

$$\varphi(\gamma) \equiv \varphi(x, y) = px + qy \; , \; p \geq 0, q \geq 0, \; p + q > 0 \; .$$

Lemma 3.4.2 *Let* $p, q \geq 0$, $p + q > 0$, *be such that the vectors* M' *and* M'' *have the following properties (see figure 3.4.2)*

$$\begin{cases} \varphi(M') = pM_x' + qM_y' \geq 0 \; , \\ \varphi(M'') = pM_x'' + qM_y'' \geq 0 \; . \end{cases} \quad (3.48)$$

Then, for all $(x, y) \in \mathbf{Z}_+^2, (x, y) \neq (0, 0)$,

$$E[\varphi(\xi_{n+1}) / \xi_n = (x, y)] \geq \varphi(x, y) \; , \quad (3.49)$$

i.e. $\varphi(\xi_n)$ *is a positive submartingale.*

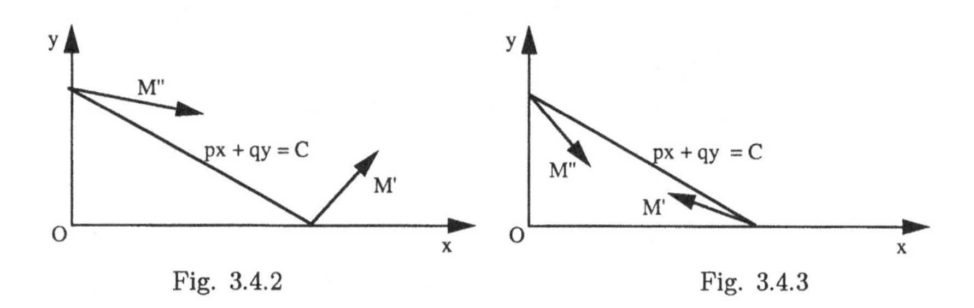

Fig. 3.4.2 Fig. 3.4.3

Proof Immediate by using the linearity of φ and the fact that $(M_x, M_y) = (0,0)$, for $x, y > 0$. ■

It is simple to check that the conditions of lemma 3.4.2 hold in the case (i) of theorem 3.4.1. They are also valid when

$$\phi_x < \frac{\pi}{2},\ \phi_y < \frac{\pi}{2},\ \phi_x + \phi_y \geq \frac{\pi}{2},$$

which yields

$$M'_x M''_y - M'_y M''_x \leq 0\,. \tag{3.50}$$

Lemma 3.4.3 *If the random walk $\mathcal{L}$ satisfies the conditions of lemma 3.4.2, then $\mathcal{L}$ is not ergodic.*

Proof This is a direct consequence of theorem 2.2.6. It is worth mentioning that this result holds under the mere assumption, weaker than condition B,

$$E(\| \theta_{n+1} \| / \xi_n = (x,y)) \leq C < \infty\,. \tag{3.51}$$

Consider now the case $\phi_x + \phi_y < \pi/2$.
Then we have the property opposite to that of lemma 3.4.2, since the vectors M' and M'' point now toward the interior of the simplex bounded by the two positive axes and the line $px + qy = C$. (see figure 3.4.3). This means that there exist $p > 0$ and $q > 0$, such that the linear function $\varphi(\xi_n)$ is now a positive supermartingale. Hence, the random walk $\mathcal{L}$ is recurrent. Our goal is to distinguish between positive and null recurrence. To that end, we introduce again the quadratic form

$$Q(x,y) = (ux^2 + vy^2 + xy)\ ,\ (x,y) \in \mathbf{Z}_+^2\ ,$$

where u and v are positive.
As in the preceding section, the game consists in adjusting u and v to satisfy the conditions of theorem 2.2.8. Define, for the sake of brevity,

$$\begin{aligned}\Delta Q(x,y) &= Q(x+\theta_x, y+\theta_y) - Q(x,y) \\ &= u\theta_x^2 + v\theta_y^2 + \theta_x\theta_y + (y+2ux)\theta_x + (x+2vy)\theta_y \,.\end{aligned} \tag{3.52}$$

Put

$$K(x,y) \stackrel{\text{def}}{\equiv} E[Q(\xi_{n+1}) - Q(\xi_n)/\xi_n = (x,y)] \,.$$

Then

$$K(x,y) = xE[2u\theta_x + \theta_y] + yE[2v\theta_y + \theta_x] + E[Q(\theta_x,\theta_y)] \,. \tag{3.53}$$

Since $E[Q(\theta_x,\theta_y)] = O(1)$, $\forall (x,y) \in \mathbf{Z}_+^2$, we get from (3.53), after taking into account the boundary conditions on the axes,

$$K(x,y) = \begin{cases} \lambda_x u + \lambda_y v + R \,, & (x,y) > 0 \,, \\ y(2vM_y'' + M_x'') + O(1) \,, & x=0,\ y>0 \,, \\ x(2uM_x' + M_y') + O(1) \,, & x>0,\ y=0 \,. \end{cases}$$

Thus, for some $\epsilon > 0$ and some finite subset $E \in \mathbf{Z}_+^2$, we have

$$K(x,y) < -\epsilon, \ \forall (x,y) \notin E \,,$$

provided that the following system can be satisfied, for some $u, v > 0$,:

$$\begin{cases} \lambda_x u + \lambda_y v + R & < 0 \,, \\ 2uM_x' + M_y' & < 0 \,, \\ 2vM_y'' + M_x'' & < 0 \,. \end{cases} \tag{3.54}$$

The inequalities $M_x'' \ge 0,\ M_y'' < 0,\ M_y' \ge 0,\ M_x' < 0$, show at once that (3.54) can be satisfied if (3.46) holds, that is

$$-\lambda_x \frac{M_y'}{M_x'} - \lambda_y \frac{M_x''}{M_y''} + 2R < 0 \,.$$

Then the remaining conditions of Foster's criterion (theorem 2.2.3) are clearly fulfilled and the random walk is ergodic.
In the other case, when (3.47) holds, i.e.

$$-\lambda_x \frac{M_y'}{M_x'} - \lambda_y \frac{M_x''}{M_y''} + 2R > 0 \,,$$

one can find u, $v > 0$, so that

$$\begin{cases} \lambda_x u + \lambda_y v + R & > 0 , \\ 2uM'_x + M'_y & > 0 , \\ 2vM''_y + M''_x & > 0 . \end{cases} \tag{3.55}$$

Hence, for some finite subset $E \subset \mathbf{Z}_+^2$, we have $K(x,y) > 0$, for $(x,y) \notin E$. Consequently, when (3.47) is satisfied, there exist two functions, namely the quadratic form $Q(x,y)$ and the linear function $\varphi(x,y)$, such that, assuming $\varphi_x + \varphi_y < \frac{\pi}{2}$, all the conditions of theorem 2.2.8 hold (taking $\alpha = 2$ in the statement of this theorem). This shows the *null recurrence* part of theorem 3.4.1, which is, by the way, completely proved. ■

The general conditions allowing for separation between transience and recurrence have been obtained in [AFM]. We simply quote them here, without proof:

Theorem 3.4.4 *The random walk is*

(i) *recurrent if*

$$-\lambda_x \frac{M''_y}{M''_x} - \lambda_y \frac{M'_x}{M'_y} + 2R \geq 0 ;$$

(ii) *transient if*

$$-\lambda_x \frac{M''_y}{M''_x} - \lambda_y \frac{M'_x}{M'_y} + 2R < 0 .$$

Problems

(1) Find the complete classification in the case

$$\lambda_x \frac{M'_y}{M'_x} + \lambda_y \frac{M''_x}{M''_y} - 2R = 0 .$$

(2) Classify random walks in $\mathbf{Z}_+^3$, when $M_{\{1,2,3\}} = 0$. In particular

(i) Are there *ergodic* cases?

(ii) When do the one-dimensional faces play no role in the classification?

3.5 Jackson networks

Jackson networks are now classical models for communication networks. Jackson [Jac63] obtained the famous product form for their stationary probabilities. Sufficient ergodicity conditions follow from this product form. The proof of the necessity of these conditions was obtained by many authors [Afa87, Bor86, Fos89, Pod85]. From the general theory of countable Markov chains, it follows then that the n-step transition probabilities converge to stationary probabilities when $n \to \infty$. But not much was known about the rate of this convergence. The only result is exponential convergence under some *smallness* assumptions in [KKR88]. Here we consider basic Markovian Jackson networks, i.e. with Poisson arrivals and exponential service times. In this case, these systems are equivalent to a class of random walks in $\mathbf{Z}_+^N$, where N is the number of nodes in a network. The main result of this section is that we give an explicit construction of Lyapounov functions. They are either *almost linear* in the terminology of [Mal72a], or just *piecewise linear.* Later on, in chapter 7, we again use this construction to show exponential convergence to the steady state (whenever it exits) and also analyticity results. Let us emphasize that we never use Jackson's product form in the proofs.

Ergodicity conditions for Jackson networks

Here we recall some well known facts and prove a useful geometric lemma. We consider an open Jackson network with N nodes. Let $\xi^i(t)$ be the length of the queue at the i-th node at time t. We restrict ourselves here to the simplest assumptions: independent Poisson inputs with parameter $\lambda_i > 0$ for any node i, exponential service times with parameters $\mu_i > 0$ and FIFO service discipline. After a customer completes service at the i-th node, he is immediately transferred with probability p_{ij} to the end of the queue at node j, $j = 1, \ldots, N$, and, with probability

$$p_{i0} = 1 - \sum_{j=1}^{n} p_{ij} \,,$$

he leaves the network. It will be convenient (although strictly not necessary) to assume $p_{ii} = 0$, for all i.
In other words, we consider a continuous-time random walk $\tilde{L}$ on $\mathbf{Z}_+^N$ with transition intensities $\lambda_{\alpha\beta}$, from the state $\alpha = (\alpha^1, \ldots, \alpha^N)$ to the

state $\beta = (\beta^1, \dots, \beta^N)$, where

$$\lambda_{\alpha\beta} = \begin{cases} \mu_{oi} = \lambda_i \,, & \text{if } \beta - \alpha = e_i, \\ \mu_{io} = \mu_i \, p_{i0} \,, & \text{if } \beta - \alpha = -e_i \\ \mu_{ij} = \mu_i \, p_{ij} \,, & \text{if } \beta - \alpha = -e_i + e_j \,, \text{ for } 1 \le i,j \le N \,. \end{cases} \tag{3.56}$$

Here e_i denotes the vector $(0, \dots, 0, 1, 0, \dots, 0)$, having its i-th coordinate equal to 1. It is convenient to denote the zero vector by e_0. We recall now Jackson's equations. Assuming a stationary regime, we denote by ν_j the mean number of customers visiting node j and coming from the outside world or from the other nodes during a unit time interval. Using the law of large numbers, Jackson wrote the following system of equations (we call it *Jackson's system*):

$$\nu_j = \lambda_j + \sum_{i=1}^{N} \nu_i \, p_{ij} \,, \quad j = 1, \dots, N \,. \tag{3.57}$$

Let us note that these equations can be solved by the iteration scheme

$$\nu_j = \lambda_j + \sum_{k=1}^{\infty} \sum_{i=1}^{N} \lambda_i \, p_{ij}^{(k)} \,, \tag{3.58}$$

where

$$\| p_{ij}^{(k)} \| = \mathbf{P}^k \,, \ \mathbf{P} = \| p_{ij} \|_{i,j=0,1,\dots,N} \,,$$

and we put $p_{oi} \equiv 0, \ i \neq 0, \ p_{oo} = 1$.
The series in the right-hand side of (3.58) converges if

$$p_{ij}^{(k)} \le C(1 - \epsilon)^k \,, \tag{3.59}$$

for some $\epsilon > 0, \ C > 0$. For this, it is necessary and sufficient to assume the classical

Condition J : *Starting from any state, the Markov chain with $N+1$ states $0, 1, \dots, N$, defined by the stochastic matrix* $\mathbf{P}$, reaches 0 with a positive probability (a.s.).

Thus, we can rewrite (3.58) as

$$\nu_j = \lambda_j + \sum_{i=1}^{N} \lambda_i \, m_{ij}^0 \,,$$

where m_{ij}^0 is the mean number of hittings of j, starting from i, in this finite-state Markov chain. Then, it is immediate to see that the solution of (3.57) is unique.

Theorem 3.5.1 *(Jackson). The network is ergodic if, and only if,*

$$\nu_j < \mu_j \text{ for all } j = 1, \ldots, N .$$

We will give a new proof of this theorem by means of a geometrical approach, which is the key point of the study. Later on we also make use of several results for discrete time Markov chains, borrowed from chapter 2. It is worth noting that all of them could be easily rewritten for the continuous time case. To avoid this rewriting, we introduce the following discrete time random walk L in $\mathbf{Z}_+^N$. Its transition probabilities are taken to be

$$p_{\alpha\beta} = w_\alpha \, \lambda_{\alpha\beta} , \tag{3.60}$$

for some constants w_α satisfying

$$0 < w_\alpha \leq (\sum_\beta \lambda_{\alpha\beta})^{-1}.$$

For instance, the choice

$$w_\alpha = (\sum_\beta \lambda_{\alpha\beta})^{-1}$$

yields the natural embedded chain. In fact, it will be more convenient to choose

$$w_\alpha \equiv w \leq \min_\alpha (\sum_\beta \lambda_{\alpha\beta})^{-1}. \tag{3.61}$$

The stationary probabilities $\tilde{\pi}_\alpha$ of $\tilde{L}_\alpha$ and those π_α of L are the same, i.e.

$$\pi_\alpha = \tilde{\pi}_\alpha, \tag{3.62}$$

so that $\tilde{L}$ is ergodic if, and only if, L is ergodic.

We shall consider the discrete time homogeneous Markov chain L, which is assumed to be irreducible and aperiodic, unless otherwise stated. The notation of section 3.2 will be in force. In particular, we recall the definition of a *face*: For any $\wedge \subseteq \{1, 2, \ldots, N\}$, the *face* $B^\wedge$ of $\mathbf{R}_+^N$ is the set

$$B^\wedge = \{(r_1, \ldots, r_N) : r_i > 0, \ i \in \wedge \, ; \ r_i = 0, \ i \notin \wedge\} .$$

Obviously the random walk L, which is equivalent to the Jackson network under study, does meet conditions A_0 and A_1 of section 3.2. Here condition A_1 is even stronger, since

$$p_{\alpha\beta} = 0 , \quad \text{for} \quad \| \alpha - \beta \| > 1 .$$

The *first vector field* on $\mathbf{R}_+^N$ is constant on any B^Λ and is equal to

$$M_\Lambda \equiv M(\alpha) \, , \; \alpha \in \Lambda \, .$$

For the Markov chain L, we have the crucial property

$$M_\Lambda = f_0 + \sum_{i \in \Lambda} f_i \, , \tag{3.63}$$

where

$$f_i = w \sum_{j=0}^{N} \mu_{ij}(-e_i + e_j) \, , \; \text{for} \;\; i = 0, \ldots, N \, . \tag{3.64}$$

So, f_i represents the contribution of the transition from the i-th node (including the virtual 0-node). On the other hand, it is clear that the 2^N mean jump vectors M_Λ are the vertices of a parallelepiped which we denote by $\partial\Pi$. Its initial point can be taken as f_0 and the edges drawn from this point are $f_1, \ldots, f_N$. This parallelepiped may be degenerate if the vectors $f_1, \ldots, f_N$ are linearly dependent. We shall use below the following combinatorial criterion of ergodicity, equivalent to Jackson's one.

Lemma 3.5.2 *The Jackson network is ergodic if, and only if, $\partial\Pi$ is not degenerate and the point $0 \in \mathbf{R}^N$ is one of its interior points. Moreover, if the origin does not belong to $\partial\Pi$, then this chain is transient.*

Proof Consider the following system of equations, with respect to $\epsilon_1, \ldots, \epsilon_N$:

$$f_0 + \epsilon_1 \, f_1 + \cdots + \epsilon_N \, f_N = 0 \, . \tag{3.65}$$

Note that $\partial\Pi$ is not degenerate if, and only if, this system is not degenerate. In this case, (3.65) has a unique solution and 0 is an internal point of $\partial\Pi$ if, and only if, $0 < \epsilon_i < 1$, for $i = 1, \ldots, N$. Inserting (3.56), (3.64) into (3.65), we get

$$\begin{aligned}
0 = f_0 + \sum_{j=1}^{N} \epsilon_j f_j &= \sum_{j=1}^{N} \lambda_j e_j + \sum_{i=1}^{N} \epsilon_i \sum_{j=0}^{N} \mu_i p_{ij}(-e_i + e_j) \\
&= \sum_{j=1}^{N} \lambda_j e_j + \sum_{i=1}^{N} \epsilon_i \mu_i [-e_i + \sum_{j=0}^{N} p_{ij} e_j] \\
&= \sum_{j=1}^{N} \lambda_j e_j - \sum_{j=1}^{N} \epsilon_j \mu_j \, e_j + \sum_{i=1}^{N} \epsilon_i \mu_i \sum_{j=0}^{N} p_{ij} \, e_j
\end{aligned}$$

$$= \sum_{j=1}^{N}(\lambda_j - \epsilon_j\,\mu_j + \sum_{i=1}^{N}\epsilon_i\,\mu_i\,p_{ij})e_j\ ,$$

which coincides with (3.57) for $\epsilon_i \equiv \rho_i = \dfrac{\nu_i}{\mu_i}$.

Thus, when 0 is an interior point of $\partial\Pi$ and $\partial\Pi$ is not degenerate, the ergodicity follows directly from Jackson's explicit formulas for the stationary probabilities *but later we shall prove it without using Jackson's results.*

Let now 0 lie on the boundary of $\partial\Pi$ (this includes the case of a degenerate $\partial\Pi$, when $\partial\Pi$ coincides with its boundary). Assume first that $\partial\Pi$ is not degenerate. Then there exists a hyperplane $\mathcal{L}$ of dimension $N-1$ in $\mathbf{R}^N$, such that $0 \in \mathcal{L}$ and $\partial\Pi$ belongs to the closure of one of the two half-spaces defined by $\mathcal{L}$. Denote this closure by $\mathcal{L}^+$ and consider the straight line l, passing through 0 and perpendicular to $\mathcal{L}$. Let x be the coordinate on l which is positive on $\mathcal{L}^+$. For any point $\alpha \in \mathbf{R}_+^N$, let $f(\alpha)$ be the value of the x-coordinate of the orthogonal projection of α onto l. Then, since all the M_Λ's belong to $\mathcal{L}^+$, it follows that

$$\sum_\beta p_{\alpha\beta} f(\beta) - f(\alpha) \geq 0\ ,\ f(\alpha) > 0 \text{ for an infinite number of } \alpha \in \mathbf{Z}_+^N .$$

Consider the sequence of random variables $\xi_0, \xi_1, \ldots$ constituting the chain L, and the corresponding sequence $f(\xi_t)$. Let τ be the time of first visit of ξ_t to the set $\{\alpha : f(\alpha) \leq 0\}$ and $\eta(t) = f(\xi_{t\wedge\tau})$. We have

$$E[\eta_{t+1}/\xi_t, \ldots, \xi_0] - \eta_t \geq 0\ .$$

It is then well known that $E\tau = \infty$, which proves the non-ergodicity by using theorem 2.1.3.

For a degenerate $\partial\Pi$, the proof is the same and, with regard to the transience, we have, for some $\epsilon > 0$,

$$\sum_\beta p_{\alpha\beta} f(\beta) - f(\alpha) \geq \epsilon\ ,\ f(\alpha) > 0 \ \text{ for an infinite number of } \alpha \in \mathbf{Z}_+^N\ .$$

Then the results follows from theorems 2.1.9 and 2.2.7. The lemma 3.5.2 is proved. ■

Geometric construction

Let us recall that $\partial\Pi$ is the convex hull of the *points* M_Λ (the ends of

the vectors $M_\wedge$ having their origin at 0).
Let a be a fixed point of $\mathbf{R}^N$. We define

$$\Gamma \equiv \Gamma^a = \{a + \sum_{i=1}^{N} \beta_i \, f_i : \beta_i \geq 0\} \, .$$

Hence, Γ represents a multidimensional cone (with vertex a) generated by the vectors f_i. It will be convenient to put

$$\Gamma_\wedge^a = \{a + \sum_{i \in \wedge} \beta_i \, f_i : \beta_i \geq 0\} \, , \wedge \subset \{1, \ldots, N\} \, ,$$

so that

$$\Gamma^a \equiv \Gamma^a_{\{1,\ldots,N\}} \, .$$

We define the *surface* $\tilde{\Gamma}^a$ of Γ^a by

$$\tilde{\Gamma}^a = \bigcup_{\wedge \neq \{1,\ldots,N\}} \Gamma^a_\wedge \, .$$

Whenever $a = f_0$, we shall simply write Γ, $\Gamma_\wedge, \tilde{\Gamma}_\wedge$, etc.

Scaling : We shall denote by $\alpha\Gamma$, $\alpha\tilde{\Gamma}$, $\alpha\Gamma_\wedge$, $\alpha \geq 1$, the respective scaled geometrical objects, with vertex αa.

Lemma 3.5.3 *The set*

$$\mathbf{R}_+^N \cap (\alpha\Gamma)$$

is compact for any $\alpha \geq 1$.

Proof Let us first note that, if this set is compact for some a, then it is compact for any a. Hence we can choose a in a convenient way, putting for instance

$$a = f_0 \, .$$

We now remark that the ray $f_0 + \beta_i f_i$, $0 \leq \beta_i < \infty$, intersects the face $x_i = 0$ of $\mathbf{R}_+^N$, since f_i has its e_i-component negative and the others are positive. From this, the required compactness is readily seen. In fact, we can rewrite

$$f_0 + \sum_{i=1}^{N} \beta_i \, f_i = w \sum_{i=1}^{N} C_i \, e_i \, , \tag{3.66}$$

with

$$C_j = \lambda_j \, + \sum_{i=1}^{N} \alpha_i \, p_{ij} - \alpha_j \, ,$$

after having set

$$\alpha_j = \mu_j \, \beta_j \, .$$

Choosing now a ray $\alpha_i = tr_i$, $r_i \geq 0$, $t \geq 0$, we see that its intersection with $\mathbf{R}_+^N$ is an interval of finite length, since

$$\sum_{j=1}^{N} C_j = \sum \lambda_j + t\left(\sum r_i(1-p_{io}) - \sum r_i\right) = \sum \lambda_j - t\sum r_i p_{io} \, ,$$

and the coefficient of t is negative. The lemma is proved ■

Let us now consider some $\Gamma_\wedge$, with $|\wedge| = N-1$. This defines an affine hyperplane $H_\wedge$ (of dimension $N-1$) in $\mathbf{R}^N$, which subdivides $\mathbf{R}^N$ into 2 half-spaces $\Gamma_\wedge^+, \Gamma_\wedge^-$. We denote by $\Gamma_\wedge^+$ the half-space containing Γ.

Lemma 3.5.4 *We assume that 0 lies inside $\partial\Pi$ and consider an arbitrary $\Gamma_\wedge$ with $|\wedge| = N-1$. Then any vector $M_{\wedge'}$, such that*

$$\wedge' \not\subset \wedge \, , \tag{3.67}$$

which has its initial point in $\Gamma_\wedge$, lies entirely in $\Gamma_\wedge^+$.

Proof Let us first show that $M_{\{1,\dots,N\}}$ has this property for all $\Gamma_\wedge$, such that $|\wedge| = N-1$. For this, choose $a = -M_{\{1,\dots,N\}}$. Let us note that $0 \in \partial\Pi$ if, and only if, $0 \in \{-\partial\Pi\}$. Then the vector $M_{\{1,\dots,N\}}$, with initial point a (which belongs to all $\Gamma_\wedge^a$ simultaneously), has 0 as its final point and is thus contained in $\Gamma_\wedge^+$.
Take now e.g. $\wedge = \{2,\dots,N\}$ and any $\wedge'$ such that $1 \in \wedge'$. Choosing again

$$a = -M_{\{1,\dots,N\}} = -f_0 - \sum_{i\in\wedge'} f_i - \sum_{i\neq 0, i\notin\wedge'} f_i \, ,$$

we see that the point

$$b = a + \sum_{i\neq 0, i\notin\wedge'} f_i$$

belongs to $\Gamma_\wedge$ and $b + M_{\wedge'} = 0 \in \Gamma_\wedge^+$. The lemma is proved ■

Lemma 3.5.5 *If a lies strictly inside $\mathbf{R}_+^N$, then $\Gamma_\wedge^a$, for any $|\wedge| = N-1$, has the property*

$$\Gamma_\wedge^a \cap \overline{B^{\wedge'}} = \partial\emptyset \quad , \quad \mathit{for}\ \wedge' \subset \wedge \, ,$$

where $\overline{B}$ denotes the closure of B.

Proof Let again $\wedge = \{2, \ldots, N\}$. Then

$$\Gamma_{\wedge}^{a} = \{a + \sum_{j=2}^{N} \beta_j \, f_j\} \, .$$

But the vector

$$a + \sum_{j=2}^{N} \beta_j \, f_j = \sum_{i=1}^{N} C_i \, e_i \, ,$$

since C_1 is strictly positive, cannot belong to the region $\overline{B^{\wedge'}}$, where the first coordinate is zero. The lemma is proved. ■

Let us introduce now the following function, with domain $\mathbf{R}_+^N$:

$$f : \, x \rightarrow f_x = \alpha, \quad \text{for} \quad x \in \alpha\tilde{\Gamma} \, . \tag{3.68}$$

This piecewise linear function, obtained by scaling, is our main Lyapounov function, as will be shown below. We shall in fact propose two constructions, which are both useful for future generalizations.

First construction

This construction uses *smoothing*, which is in fact equivalent to the *principle of almost linearity*, presented in section 3.3.

Lemma 3.5.6 *For any $\epsilon > 0$, there exists a smooth convex closed hypersurface $\partial\Pi(\epsilon)$, homeomorphic to the boundary $\partial\Pi$ of Π such that, for any $x \in \partial\Pi(\epsilon)$,*

$$\rho(x, \partial\Pi) < \epsilon \, ,$$

and, for any $y \in \partial\Pi$,

$$\rho(y, \partial\Pi(\epsilon)) < \epsilon \, ,$$

ρ being an arbitrary metric.

For the proof, it is sufficient to consider the unit cube in $\mathbf{R}^N$ and then to use a linear transformation. For this unit cube, one can proceed by induction, in constructing at each step a cylinder *smoothed* at the ends. Take the intersection of $\partial\Pi(\epsilon)$ with a neighbourhood of the vertex $a = f_0$. We extend it by linearity to get a hypersurface $\tilde{\Gamma}(\epsilon)$, smooth and convex, such that the pairs

$$(\tilde{\Gamma} \cap \mathbf{R}_+^N, \, \tilde{\Gamma} \cap \partial(\mathbf{R}_+^N))$$

and

$$(\tilde{\Gamma}(\epsilon) \cap \mathbf{R}_+^N,\ \tilde{\Gamma}(\epsilon) \cap \partial(\mathbf{R}_+^N))$$

are homeomorphic. Moreover,

$$\begin{aligned} \rho(x, \tilde{\Gamma}(\epsilon)) &< \epsilon, \quad \text{for any } x \in \tilde{\Gamma} \cap \mathbf{R}_+^N , \\ \rho(y, \tilde{\Gamma}) &< \epsilon, \quad \text{for any } y \in \tilde{\Gamma}(\epsilon) \cap \mathbf{R}_+^N . \end{aligned}$$

Then, by scaling, we define the following Lyapounov function:

$$f : x \to f_x = \alpha \ , \quad x \in \alpha\tilde{\Gamma}(\epsilon) . \tag{3.69}$$

Let us fix some $r > 0$ sufficiently large and an arbitrary $x \in \alpha\tilde{\Gamma}(\epsilon)$. Let $O(x) = O(r; x)$ be a cube with sides of length r, centred at x. Then there exist linear functions $g_x(\cdot)$ on $\mathbf{R}^N$ such that

$$\sup_{x \in \alpha\tilde{\Gamma}(\epsilon)} \ \sup_{y \in O(x)} \mid f_y - g_x(y) \mid \to 0 \ \text{ for } \alpha \to \infty .$$

By geometric construction, when 0 lies inside Π, we have, for all sufficiently large $x \in Z_+^N$ and for some $\delta > 0$,

$$f_{x+M(x)} - f_x < -\delta .$$

From the last two formulas, it follows that, for some $\delta_1 > 0$ and for all x except for a finite set, we have

$$\sum_y p_{xy} f_y - f_x < -\delta_1 .$$

This yields the following (as announced, we have used the principle of almost-linearity)

Theorem 3.5.7 *If* O *lies inside* $\partial\Pi$, *then the function* f, *defined by (3.69), satisfies the conditions of theorem 2.2.3, so that the Jackson network is ergodic.*

Second construction

In this construction, we shall use the Lyapounov function (3.68) together with theorem 2.2.4, for integer-valued functions $k(x) \equiv k$, with k taken sufficiently large

Theorem 3.5.8 *Let us consider a Jackson network such that* $0 \in \partial\Pi$, *and choose the function* f_x *as in (3.68), with the point* a *lying inside* $\mathbf{R}_+^N$. *Then, for this Lyapounov function, and* $k(x) = k$ *sufficiently large,* *the conditions of theorem 2.2.4 are satisfied and the Jackson network is ergodic.*

Proof Let us fix a constant d, representing the maximal length of a jump, for some metric ρ in $\mathbf{R}^N$. In our case, $d = 1$ for the metric

$$\rho(x, y) = \max_i |x_i - y_i| ,$$

where e.g. $x = (x^1, ..., x^N)$.

Lemma 3.5.9 *Let us fix $\epsilon > 0$ sufficiently small. There exists $\rho_0 > 0$ such that, for any $x \in \mathbf{Z}_+^N$, with*

$$\rho \equiv \rho(x, \partial \mathbf{R}_+^N) > \rho_0 ,$$

and for any k such that

$$\rho_0 < kd \equiv k \leq \rho ,$$

we have

$$\sum_y p_{xy}^{(k)} f_y - f_x < -\epsilon . \tag{3.70}$$

Proof For any $\wedge$, $|\wedge| = N-1$, let us consider a new function $f^\wedge : x \to f_x^\wedge = \alpha$, for x belonging to the hyperplane $\alpha H_\wedge$ generated by $\alpha \Gamma_\wedge$. Consider the evolution of the random walk after k steps, assuming it started from x. That is

$$x = \xi_0,\ \xi_1, \ldots, \xi_{k-1} .$$

Then the sequence

$$f_{\xi_0}^\wedge,\ f_{\xi_1}^\wedge, \ldots, f_{\xi_{k-1}}^\wedge$$

is a supermartingale satisfying, for some $\epsilon' > 0$,

$$E(f_{\xi_i}^\wedge / f_{\xi_0}^\wedge, \ldots, f_{\xi_{i-1}}^\wedge) - f_{\xi_{i-1}}^\wedge < -\epsilon' ,$$

since $M_{\{1,\ldots,N\}}$ is directed to the corresponding side of the hyperplane $\alpha H_\wedge$. Thus, by theorem 2.1.7, we can find constants $C_\wedge$, $\delta_\wedge$, $\epsilon_\wedge > 0$, such that

$$f_{\xi_{k-1}}^\wedge - f_x^\wedge < -k\epsilon_\wedge , \tag{3.71}$$

with probability $1 - C_\wedge\, e^{-k\delta_\wedge}$. Hence, for ρ_0 chosen sufficiently large, (3.71) takes place for all $\wedge$, with probability

$$1 - C\, e^{-\delta k} ,$$

for some constants $C,\ \delta > 0$. Using the boundedness of jumps and the fact that the function f grows linearly with α, we immediately get (3.70), for some $\epsilon > 0$.
The lemma is proved. ∎

Lemma 3.5.10 *Again choose $\epsilon > 0$ sufficiently small and $i \in 1, \ldots, N$. Then there exist $\rho_i > 0$ such that, for any $x \in \mathbf{Z}_+^N$ with*

$$\rho \equiv \max_{\wedge : i \notin \wedge} \rho(x, B^{\wedge}) > \rho_i \ , \tag{3.72}$$

and for any k such that

$$\rho_i < kd \equiv k \leq \rho \ ,$$

we have

$$\sum_y p_{xy}^{(k)} f_y - f_x < -\epsilon \ . \tag{3.73}$$

Proof We repeat the proof of lemma 3.5.9, using the geometrical construction, as the mean jump vectors point in a suitable direction, from any point which the random walk, starting from x, can possibly visit during $\rho - 1$ steps. The lemma is proved. ∎

To prove the theorem, we introduce the quantity

$$\tilde{\rho} = \max_{0 \leq i \leq N} \rho_i \ .$$

Then, for any point x lying outside a $(\tilde{\rho}+1)$-neighbourhood of the origin, in our special metric, we put

$$k(x) \equiv k \equiv \tilde{\rho}$$

and note that, for any such point, there exists i such that, for $\wedge = \{1, \ldots, N\} - \{i\}$, inequalities (3.72) and (3.73) hold. The proof of theorem 3.5.8 is concluded. ∎

3.6 Asymptotically small drifts

Let us consider a discrete time MC $\{X_n\}$, with state space $\mathbf{Z}_+$ and satisfying the following moment condition:

$$\sup_{n \geq 0} E[\mid X_{n+1} - X_n \mid^{2+\epsilon} / X_n] < c < \infty, \ \text{ a.s., for some } \epsilon > 0, c > 0 \tag{3.74}$$

Let us define for $x \in \mathbf{Z}_+$ and $i = 1, 2, \ldots$

$$\mu_i(x) = E[(X_{n+1} - X_n)^i / X_n = x] \tag{3.75}$$

From the condition (3.74), it follows that

$$\sup_{x \in \mathbf{Z}_+} \mu_2(x) < \infty \quad \text{and} \quad \sup_{x \in \mathbf{Z}_+} \mu_1(x) < \infty .$$

The next theorem, proved in [AIM], generalizes results obtained by Lamperti [Lam60].

Theorem 3.6.1 *The following classification holds:*

(i) *If there exists a number B such that*

$$\mu_1(x) \le \frac{\mu_2(x)}{2x} \quad , \text{ for } x \ge B , \tag{3.76}$$

then $\{X_n\}$ is recurrent.

(ii) *If instead, for some $\theta > 1$,*

$$\mu_1(x) \ge \frac{\theta \mu_2(x)}{2x} \quad \text{for } x \ge B , \tag{3.77}$$

then $\{X_n\}$ is transient.

(iii) *If*

$$\mu_1(x) \ge -\frac{\mu_2(x)}{2x} \quad , \text{ for } x \ge B , \tag{3.78}$$

then $\{X_n\}$ is non-ergodic.

(iv) *If for some $\theta > 1$*

$$\mu_1(x) < -\frac{\theta \mu_2(x)}{2x} \quad , \text{ for } x \ge B , \tag{3.79}$$

then $\{X_n\}$ is ergodic.

Remark The transience and recurrence parts of this theorem are due to Lamperti [Lam60]. Non-ergodicity and ergodicity can be proved using theorem 2.2.8. These results, and also the following one (more precise), are stated without the proofs, which can be found in [AIM].

We define recursively

$$\log x = \log^{(1)}(x), \ \log(\log x) = \log^{(2)} x, \ \ldots, \log^{(k)}(x) = \log(\log^{(k-1)} x), \ldots \text{etc} .$$

Theorem 3.6.2

(i) *If for some $k \in Z_+$ there exists a number B such that, for $x \geq B$,*

$$\mu_1(x) \geq -\frac{\mu_2(x)}{2x} - \frac{\mu_2(x)}{2x \log x} - \frac{\mu_2(x)}{2x \log x \log^{(2)}(x)} - \cdots - \frac{\mu_2(x)}{2x \prod_{p=1}^{k} \log^{(p)}(x)}, \tag{3.80}$$

then $\{X_n\}$ is non-ergodic.

(ii) *If, for some k, for some $\theta > 1$ and for $x \geq B$,*

$$\mu_1(x) < -\frac{\mu_2(x)}{2x} - \frac{\mu_2(x)}{2x \log x} - \cdots - \frac{\mu_2(x)}{2x \prod_{p=1}^{k-1} \log^{(p)}(x)} - \frac{\theta \mu_2(x)}{2x \prod_{p=1}^{k} \log^{(p)}(x)}, \tag{3.81}$$

then $\{X_n\}$ is ergodic.

Theorem 3.6.1 could be used as a classification criterion in multidimensional zero drift situations. The algorithm might be as follows: find a family of Lyapounov functions $f(\vec{a}, \alpha)$, where $\alpha \in \mathcal{A}$ is the state space of the MC and $\vec{a}$ is a real vector parameter, satisfying two conditions:

(i) $\vec{a}$ is small and for $\vec{a} = 0$ f is a linear function;

(ii) $f(\vec{a}; \alpha)$ is a super- (or sub-) martingale with asymptotically small drift. After this we could just use the theorem above mentioned. This way might be a more direct alternative approach to section 3.4.

An example of such a theorem is the following result [Lam60], which we state now.

Let $X_n \in \mathbf{R}_+$ be a real nonnegative process (not necessarily a Markov process), satisfying

$$\begin{aligned} &E[| X_{n+1} - X_n |^{2+\epsilon} / X_{n-1}, X_{n-2}, \ldots] < C < \infty\,, \\ &\limsup_{n\to\infty} X_n = \infty\,, \quad \text{a.s.} \end{aligned} \tag{3.82}$$

We define

$$\begin{cases} \overline{\mu}(x) = \text{ess sup } E[X_{n+1} - X_n \mid X_n = x, X_{n-1}, \ldots, X_0], \\ \overline{v}(x) = \text{ess sup } E[(X_{n+1} - X_n)^2 \mid X_n = x, X_{n-1}, \ldots, X_0], \end{cases} \tag{3.83}$$

where the sup is taken over n and over the values of $\{X_i, i \leq n-1\}$. Similarly, $\underline{\mu}(x)$ and $\underline{v}(x)$ are defined by replacing sup by inf in (3.83). The finiteness of $\overline{\mu}$ and $\overline{v}$ follows from (3.83), but, as an additional assumption, $\underline{v}(x)$ is supposed to be bounded away from 0, to avoid trivial situations.

Theorem 3.6.3 *Let a non-negative stochastic process $\{X_n\}$ satisfy (3.83) and*

$$P(\limsup_{n\to\infty} X_n = \infty) = 1 \,, \tag{3.84}$$

and, for all large x,

$$\overline{\mu}(x) \leq \frac{\underline{v}(x)}{2x} + O(x^{-1-\delta}),\ \delta > 0 \,. \tag{3.85}$$

Then $\{X_n\}$ is recurrent. If instead, for some $\theta > 1$,

$$\underline{\mu}(x) \geq \frac{\theta \overline{v}(x)}{2x} \tag{3.86}$$

for all large x, then $\{X_n\}$ is transient.

Lamperti [Lam60] applied this theorem to multidimensional random walks without boundaries and with asymptotically zero (or constant) drifts. Assume that

$$\mathbf{X}_n = \begin{pmatrix} X_n^{(1)} \\ \cdot \\ \cdot \\ \cdot \\ X_n^{(s)} \end{pmatrix}, \quad n = 0, 1, 2, ...,$$

are random vectors forming a Markov process, with the transition probability function

$$F(y_1, \ldots, y_s; \mathbf{x}) = P(X_{n+1}^{(i)} - X_n^{(i)} \leq y_i,\ i = 1, \ldots, s / \mathbf{X}_n = \mathbf{x}) \,, \tag{3.87}$$

independently of n. For simplicity in the proofs, Lamperti makes the assumption that, for some $B < \infty$,

$$\mid X_{n+1}^{(i)} - X_n^{(i)} \mid \leq B \,, \text{ a.s. for } i = 1, \ldots, s \,, \tag{3.88}$$

for all n, although, as usual, (3.88) could be relaxed to a moment condition at the expense of some labor. We shall use the notation

$$\left\{ \begin{array}{lcl} E[\mathbf{X}_{n+1} - \mathbf{X}_n / X_n = \mathbf{x}] & = & \mu(\mathbf{x}) \,, \\ E\left[(\mathbf{X}_{n+1} - \mathbf{X}_n)(\mathbf{X}_{n+1} - \mathbf{X}_n)^T / \mathbf{X}_n = \mathbf{x}\right] & = & \mathbf{v}(\mathbf{x}) = \{v_{ij}(\mathbf{x})\} \,. \end{array} \right. \tag{3.89}$$

The idea is to define the process

$$R_n = \parallel \mathbf{X}_n \parallel = \{\sum_{i=1}^{s} (X_n^{(i)})^2\}^{\frac{1}{2}}$$

and apply the result of theorem 3.6.3 to $\{R_n\}$. The requirement that $\limsup R_n = \infty$ a.s. is assumed *ad hoc* (see (3.84)) and is usually easy to verify in particular cases.

Theorem 3.6.4 *Under the above hypotheses, suppose*

$$\mathbf{v}(\mathbf{x}) = \mathbf{v} + O(\| \mathbf{x} \|^{\delta}) \,, \text{ for some } \delta > 0 \,,$$

where the matrix $[v_{ij}] = v$ *is positive definite. Then* $\{\mathbf{X}_n\}$ *is recurrent provided that*

$$\mathbf{x}^T \mathbf{v}^{-1} \mu(\mathbf{x}) \leq \frac{(2-s)}{2} + O(\| \mathbf{x} \|^{-\delta})$$

for all sufficiently large $\| \mathbf{x} \|$, *while in the case for large* $\| \mathbf{x} \|$ *where*

$$\mathbf{x}^T \mathbf{v}^{-1} \mu(\mathbf{x}) \geq \frac{(2-s)}{2} + \epsilon, \ \epsilon > 0 \,,$$

then $\{X_n\}$ *is transient.*

3.7 Stability and invariance principle

Let $X^{(\theta)} = \{X_n^{(\theta)}, n \geq 0\}$ be a family of irreducible, aperiodic Markov chains, with state space $\mathbf{Z}_+$, indexed by some positive parameter θ and supposed to be ergodic for any $\theta > 0$. We shall suppose that the transition probabilities of these chains have a property of convergence

$$p_{ij}^{(\theta)} \to p_{ij} \,, \text{ as } \theta \to 0 \,,$$

where $\|p_{ij}\|$ denotes the matrix of transition probabilities corresponding to $X^{(0)}$. Let $\pi_j^{(\theta)}$ be the stationary probabilities for the matrix $\|p_{ij}^{(\theta)}\|$ and let ζ^θ denote a random variable having the distribution $P[\zeta^\theta = j] = \pi_j^{(\theta)}$. Under some assumptions explained below, we shall consider the asymptotic behaviour of the distribution of ζ^θ, as $\theta \downarrow 0$. When $X^{(0)}$ is ergodic, the problem mentioned above is generally referred to as the *stability* of the ergodic distribution of $X^{(\theta)}$. However, the discussion below deals also with the situation when $X^{(0)}$ can be *non-ergodic*. The approach proposed follows [BFK92] and could also be used for non-Markov processes.

The notation is compatible with section 3.6. In particular, for the first and second moments of the drifts, we simply add the superscript (θ).

We enforce now the following general conditions:

$$\begin{cases} \lim_{i\to\infty,\theta\downarrow 0} \mu_1^{(\theta)}(j) & = & 0 \quad \text{and} \quad \sup_{i\geq 0,\theta\geq 0} \mu_2^{(\theta)}(j) < \infty\,, \\ \lim_{i\to\infty,\theta\downarrow 0} j[\mu_1^{(\theta)}(j)+\theta] & = & -\mu\,, \quad \text{for } -\infty \leq \mu \leq +\infty\,, \\ \lim_{i\to\infty,\theta\downarrow 0} \mu_2^{(\theta)}(j) & = & b\,, \quad 0 < b < \infty\,. \end{cases} \tag{3.90}$$

Thus $X^{(\theta)}$ is ergodic, for any $\theta > 0$. We state now the following theorems, without proofs, which can be found in [BFK92].

Theorem 3.7.1 *(Ergodicity and stability). If (3.90) holds, $2\mu < b$ and*

$$\sup_{i\geq 0,\theta\geq 0} E[(X_1^{(\theta)} - i)^{2+\epsilon}/X_0^{(\theta)} = i] < C < \infty\,, \tag{3.91}$$

where ϵ is an arbitrary but strictly positive number, then the chain $X^{(0)}$ is ergodic and becomes stable, in the sense that

$$\zeta^{(\theta)} \xrightarrow{\mathcal{D}} \zeta^{(0)}, \quad as\ \theta \downarrow 0\,,$$

where $\mathcal{D}$ indicates a weak convergence. The case $\mu = \infty$ is covered by the statement of the theorem.

Theorem 3.7.2 *(Convergence to a Γ distribution). If (3.90) holds and $-\infty < 2\mu < b$, then $X^{(0)}$ is non-ergodic. If, moreover, the series representing*

$$\mu_2^{(\theta)}(j) = \sum_{k\geq -j} p_{j,j+k}^{(\theta)} k^2$$

converges uniformly with respect to j and θ, then

$$2\theta\zeta^{(\theta)} \xrightarrow{\mathcal{D}} \Gamma_{1/b,\,1-2\mu/b}\,, \quad as\ \theta \downarrow 0\,,$$

where, up to a slight abuse of langage, $\Gamma_{\alpha,\beta}$ denotes a random variable distributed according to the standard $\Gamma_{\alpha,\beta}$ distribution.

Theorem 3.7.3 *(Convergence to the uniform distribution). If (3.90), (3.91) hold, $2\mu = b$ and*

$$2j[\mu_1^{(\theta)}(j) + \theta] + \mu_2^{(\theta)}(j) = o\left(\theta + \frac{1}{j}\right)\,, \tag{3.92}$$

then $X^{(0)}$ is null recurrent and

$$\frac{\log(\zeta^{(\theta)})}{\log(1/\theta)} \xrightarrow{\mathcal{D}} U[0,1]\,, \quad as\ \theta \downarrow 0\,,$$

where $U[0,1]$ denotes the uniform distribution on $[0,1]$.

The next theorem, which gives a detailed account of the behaviour of $X^{(0)}$, is intimately related to theorems 3.6.2 and 3.6.3 and also to the studies [Lam63, Twe76].

Theorem 3.7.4 *(Distinction between transience and null recurrence). Assuming here $\theta = 0$ and that (3.90), (3.91) hold, we have the following classification:*

(i) *If $2\mu > b$, then $X^{(0)}$ is ergodic (assumption (3.91) is here not necessary).*
(ii) *If $-b \le 2\mu < b$, then $X^{(0)}$ is nullrecurrent; this is also the case if $2\mu = b$ and (3.92) holds.*
(iii) *If $b > -2\mu$, then $X^{(0)}$ is recurrent.*
(iv) *If $0 < b < -2\mu$, then $X^{(0)}$ is transient.*

It is worth making some additional remarks (see again [BFK92]):

(i) In theorem 3.7.2, the second equation of (3.90) can be replaced by the representation

$$\mu_1^{(\theta)}(j) = -\theta - \frac{\mu}{j} + o\left(\theta + \frac{1}{j}\right) , \quad \text{as } j \to \infty,\ \theta \to 0 ;$$

(ii) In the case $\mu = -\infty$, the limit distribution of $\zeta^{(\theta)}$, after a suitable normalization will be normal (see [Kor90]);
(iii) The statement of theorem 3.7.3 says, roughly speaking, that $\zeta^{(\theta)}$ is distributed as $\theta^{-\eta}$, where η is a random variable distributed as $U[0,1]$. Taking a different rate of convergence to 0 in (3.92) could yield a different distribution for $\log(\zeta^{(\theta)})$ as well.

4

Ideology of induced chains

In sections 4.1 and 4.2 we introduce the main notions (*second vector fields, paths*) and we give the preliminary description and classification of properties of these paths. In section 4.3 we get sufficient conditions for a random walk to have a Lyapounov function satisfying the simplest of Foster's criteria (with $k_i < C$, see corollary 2.2.5). In section 4.4, it is proved that these results allow for getting a complete classification in dimensions 2 and 3.

4.1 Second vector field

We come back to the terminology of section 3.2 to introduce here some fundamental concepts, frequently used in the sequel.

Definition 4.1.1 *For any $\Lambda \neq \Lambda(1, \ldots, N)$ we choose an arbitrary point $a \in B^{\Lambda} \cap \mathbf{Z}_+^N$ and draw a plane C^{Λ} of dimension $N - |\Lambda|$, perpendicular to B^{Λ} and containing a. We define the* induced *Markov chain $\mathcal{L}^{\Lambda}$, with state space $C^{\Lambda} \cap \mathbf{R}_+^N$ (by an obvious abuse in the notation, we shall write C^{Λ} most of the time) and transition probabilities*

$$
{}_{\Lambda}p_{\alpha\beta} = p_{\alpha\beta} + \sum_{\gamma \neq \beta} p_{\alpha\gamma}, \ \forall \alpha, \beta \in C^{\Lambda} ,
$$

where the summation is performed over all $\gamma \in \mathbf{Z}_+^N$, such that the straight line connecting γ and β is perpendicular to C^{Λ}. It is important to note that this construction does not depend on a.

We shall make a series of assumptions $\mathbf{0}_i$, $i = 1, 2, \ldots$. All of them hold for all points in the parameter space, except for some hypersurface of lower dimension. More exactly, we define the parameter space $\mathcal{P} \equiv \mathcal{P}_d$

depending on a fixed constant d. Due to conditions A_0 and A_1 of section 3.2, it is the product of 2^N simplices

$$\mathcal{P}(\Lambda) = \left\{ p(\Lambda;\alpha) : \ \alpha \in \mathbf{Z}_+^N, p(\Lambda;\alpha) \geq 0, \ \sum_\alpha p(\Lambda;\alpha) = 1 \right\}$$

in the finite-dimensional Euclidean space.

Assumption 0_1 For any Λ the chain $\mathcal{L}^\Lambda$ is irreducible and aperiodic. We call B^Λ (or simply Λ) *ergodic* (*non-ergodic, transient, ...*) according as $\mathcal{L}^\Lambda$ is ergodic (non-ergodic, transient,...). For an ergodic $\mathcal{L}^\Lambda$, let $\pi^\Lambda(\gamma), \gamma \in C^\Lambda$, be its stationary transition probabilities. We introduce the vector $v^\Lambda = (v_1^\Lambda, \ldots, v_N^\Lambda)$ by setting

$$\begin{aligned} v_i^\Lambda &= 0, \ i \notin \Lambda\,, \\ v_i^\Lambda &= \sum_{\gamma \in C^\Lambda} \pi^\Lambda(\gamma) M_i(\gamma), \ i \in \Lambda\,. \end{aligned}$$

Intuitively, one can imagine that the random walk starts from a point which is close to Λ, but sufficiently far from all other faces $B^{\Lambda'}$, with $\Lambda \not\subset \Lambda'$. After some time (sufficiently long, but less than the minimal distance from the above mentioned $B^{\Lambda'}$), the stationary regime in the induced chain will be established. In this regime, one can ask about the mean drift along Λ: it is defined exactly by v^Λ. For $\Lambda = \{1, \ldots, N\}$, we call Λ *ergodic*, by definition, and put

$$v^\Lambda \equiv M(\alpha), \ \alpha \in B^\Lambda \cap \mathbf{Z}_+^N\,.$$

From now on, when speaking about the *components* of v^Λ, we mean the components v_i^Λ with $i \in \Lambda$.

Assumption 0_2 $v_i^\Lambda \neq 0$, for each $i \in \Lambda$.

Definition 4.1.2 *Let us fix Λ, Λ_1, so that $\Lambda \supset \Lambda_1, \Lambda \neq \Lambda_1$, that is to say $\overline{B}^\Lambda \supset B^{\Lambda_1}$ ($\overline{B}^\Lambda$ is the closure of B^Λ). Let B^Λ be ergodic. Thus v^Λ is well defined. There are three possibilities for the direction of v^Λ w.r.t. B^{Λ_1}. We say that B^Λ is an* ingoing (outgoing) face *for B^{Λ_1}, if all the coordinates v_i^Λ for $i \in \Lambda - \Lambda_1$ are negative (positive). Otherwise we say that B^Λ is* neutral.

As an example we give simple sufficient criteria for a face to be ergodic.

Proposition 4.1.3 *A face B^Λ of dimension $N-1$ is ergodic if and only if $v_i^{\{1,\ldots,N\}} < 0$, for $i \equiv \{1, \ldots, N\} - \Lambda$.*

This statement is obvious, since in this case $\mathcal{L}^\Lambda$ is a one-dimensional Markov chain. ∎

Proposition 4.1.4 *If all faces B^Λ with $\Lambda \supset \Lambda_1$ are ergodic and ingoing for B^{Λ_1}, then B^{Λ_1} is ergodic.*

We leave the proof to the reader, since it is a direct consequence of the main result of section 4.3. ∎

Definition 4.1.5 *To any point $x \in \mathbf{R}_+^N$, we assign a vector $v(x)$ and call this function the* second vector field. *It can be multivalued on some non-ergodic faces. We put, for ergodic faces B^Λ,*

$$v(x) \equiv v^\Lambda , \ x \in B^\Lambda .$$

If B^{Λ_1} is non-ergodic, then at any point $x \in B^{\Lambda_1}$, $v(x)$ takes all values v^Λ for which B^Λ is an outgoing face with respect to B^{Λ_1}. In other words, for x belonging to non-ergodic faces, with $|x|$ sufficiently large,

$$x + v(x) \ \in \mathbf{R}_+^N ,$$

for any value $v(x)$. If there is no such vector, we put $v(x) = 0$, for $x \in B^{\Lambda_1}$. Points $x \in \mathbf{R}_+^N$, where $v(x)$ is more than single-valued, are called branch points.

There are few interesting examples, for which only the first vector field suffices to obtain ergodicity conditions for the random walk of interest, but it is nevertheless the case for Jackson networks. In general, the second vector field must be introduced. The following proposition demonstrates the usefulness of this second vector field.

Proposition 4.1.6 *If for some ergodic face Λ all components of v^Λ are positive then the random walk is transient.*

Proof Let us fix an ergodic face Λ and let ξ_t be the corresponding induced (ergodic) Markov chain. Introduce also the mutually independent random variables $\eta(t, x)$, enumerated by $x \in C^\Lambda$, $t = 0, 1, \ldots$, and defined as follows: They take values in $\mathbf{Z}^k$, $k = |\Lambda|$, so that

$$P(\eta(t, x) = y) = \sum_z p_{\alpha\beta} ,$$

where $a, y \in B^\Lambda, \alpha = (a, x), \beta = (a + y, z)$ and the summation is over all $z \in C^\Lambda$.

Let us consider now the following process with values in $\mathbf{Z}^k$, $k = |\wedge|$:

$$S_n = S_0 + \sum_{t=0}^{n-1} \eta(t, \xi_t) ,$$

where $S_0 \in B^\wedge \cap \mathbf{Z}_+^N$. This process is a random functional over the induced chain. By the ergodic theorem [KSF80], we have

$$\frac{1}{n} \sum_{t=0}^{n-1} \eta(t, \xi_t) \to v^\wedge \quad \text{a.s.}$$

Then there exist sufficiently small $\epsilon > 0$ and $\eta_0 > 0, \delta > 0$ such that, for any component $S_{n,i}$ of S_n $(i \in \wedge)$,

$$S_0 + (v_i^\wedge - \epsilon)n < S_{n,i} < S_0 + (v_i^\wedge + \epsilon)n , \tag{4.1}$$

for all $n > n_0$ with probability not less than δ.
Now take S_0 sufficiently far from the boundary of $B^\wedge$. From (4.1) we see that there exists a set A of trajectories S_n, for $n = 0, 1, \ldots$, which never reach the boundary of $B^\wedge$, $S_{n,i} \to \infty$, for all $i \in \wedge$ and, moreover, A has a positive probability.

Coming back to our random walk in $\mathbf{Z}_+^N$, we start it from the point having coordinates $S_{0,i}$, for $i \in \wedge$, and zero otherwise. Now we take the set A' of all trajectories $\omega_n, n = 0, 1, \ldots,$ of this random walk, so that the projection of A' onto $B^\wedge$ coincides with A. Noting that A and A' have the same probabilities, the transience is proved. ■

4.2 Classification of paths

We shall consider paths $\Gamma = \Gamma(t)$, that is, continuous mappings $\Gamma : [0, T] \to \mathbf{R}_+^N$, where T can be equal to ∞, such that

(i) $\Gamma(t)$ belongs to the union of ergodic faces except for some countable subset $\mathcal{F} = \mathcal{F}(\Gamma)$ of $[0, T]$;
(ii) for the points of the same interval belonging to $[0, T] - \mathcal{F}$, where the path runs through an ergodic face $B^\wedge$, $\Gamma(t)$ is linear with velocity $\frac{d\Gamma(t)}{dt} = v^\wedge$, $\Gamma(t) \in B^\wedge$.

Let us consider the increasing sequence

$$0 \le t_1 < \ldots < t_n < \ldots, \tag{4.2}$$

of times when $\Gamma(t)$ changes a face. More exactly, this sequence comprises all points t, but those for which there exists $\epsilon > 0$ such that, for

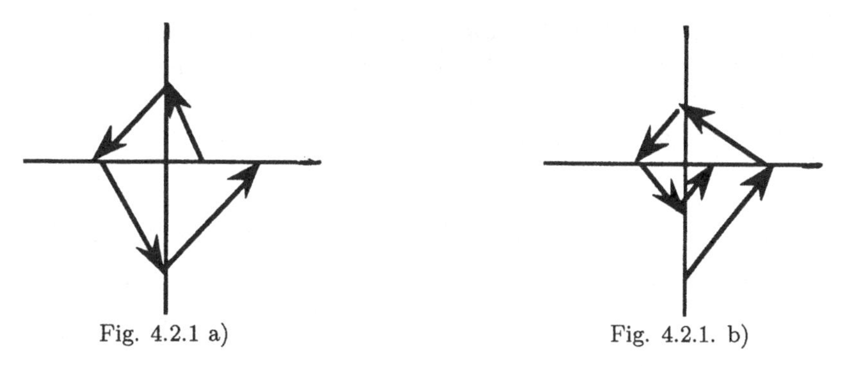

Fig. 4.2.1 a)   Fig. 4.2.1. b)

all $t' \in (t-\epsilon, t+\epsilon)$, $\Gamma(t')$ belongs to the same face as $\Gamma(t)$. The sequence (4.2) is countable and $\mathcal{F}$ belongs to it but is not equal to it because, for example, from an ergodic two-dimensional chain one can pass immediately to an ergodic one-dimensional chain. Note that the sequence (4.2) can have accumulation points, e.g. one can approach and go away from a one-dimensional face, by rotating along two-dimensional faces of $\mathbf{R}_+^N$, as shown in fig. 4.2.1, where the one-dimensional face is perpendicular to the sheet of paper at the origin. Hence, in general, the sequence (4.2) is not isomorphic to some subset of the ordered set of natural numbers, but in many examples it is (and then it is finite for finite T). Together with the sequence (4.2), to any Γ is associated a denumerable ordered set of faces

$$\Lambda_1(\Gamma), \Lambda_2(\Gamma), \dots \Lambda_n(\Gamma), \dots \tag{4.3}$$

which are visited by Γ in the corresponding order. Here $\Lambda_1(\Gamma)$ is the face to which $\Gamma(0)$ belongs. The system of all different paths starting from $x \in \mathbf{R}_+^N$ is called an *x-bundle* of paths and is denoted by V_x and the paths starting from x are denoted by Γ_x. For any visited ergodic face Λ_n, we have

$$\Gamma_x(t) = \Gamma_x(t_{n-1}) + v^{\Lambda_{n-1}}(t - t_{n-1}), \; t_{n-1} < t \le t_n .$$

Definition 4.2.1 *A point x is called* regular *if, for any $0 < T < \infty$, there exists only a finite number $n(T, V_x)$ of paths $\Gamma_x \in V_x$ on the time interval $[0,T]$. For example, $x = 0$ in fig. 4.2.1(a) is not regular since, for any $T > 0$, $n(T, V_0)$ is a continuum. For a regular point and for any Λ_i, there exists a next face in the sequence (4.3), which is denoted by Λ_{i+1}. A random walk is called* regular *if all points $x \neq 0$ are regular.*

A regular random walk is called strongly regular *if the sequence (4.2) is isomorphic to a subset of the ordered set of natural numbers for all* $x \neq 0$ *and all paths* Γ_x. *A regular point* x *is called* stable *if, for any* $\Gamma = \Gamma_x \in V_x$ *and any ergodic* $\wedge_i(\Gamma)$,

$$\dim \wedge_i (\Gamma) = \dim \wedge_{i+1} (\Gamma) + 1 ,$$

whenever $\wedge_{i+1}$ *is the next face for* $\wedge_i$. *This means that, after moving along an ergodic face* $\wedge$, *one enters a face of* $\wedge$ *of dimension* $\dim\wedge - 1$.

We show that stability implies continuity. More exactly the following proposition holds.

Proposition 4.2.2 *For any stable* x *and any* $T, \epsilon > 0$, *one can find* $\delta > 0$ *such that, for any* y *with* $\| y - x \| < \delta$, *we have*

(i) $n(T, V_x) = n(T, V_y)$;

(ii) *there exists a one-to-one correspondence* $\phi : V_x(T) \to V_y(T)$ *such that, for any* $\Gamma_x \in V_x(T)$ *and* $\Gamma_y = \phi(\Gamma_x) \in V_y(T)$, *we have*

$$|\Gamma_x(t) - \Gamma_y(t)| < \epsilon, \ t \in [0, T] .$$

Proof Let x be stable and belong to an ergodic face $\wedge$. Then for all y sufficiently close to x, Γ_x and Γ_y meet the same face (say $\wedge'$) after $\wedge$. If $\wedge'$ is non-ergodic and the second vector field on it is multivalued, then we define ϕ so that Γ_x and $\Gamma_y = \phi(\Gamma_x)$ go to the same face. Then one can proceed by induction in a finite number of steps, for any fixed finite T. The proposition is proved. ■

Now we will show that stability (and consequently continuity) is in fact a generic property for strongly regular random walks.

Proposition 4.2.3 *There exists a subset* $\mathcal{P}^0 \subset \mathcal{P}$ *of Lebesgue measure* 0 *(in fact it is the union of a finite number of analytic hypersurfaces in* $\mathcal{P}$*) such that, for any strongly regular random walk with parameters in* $\mathcal{P} - \mathcal{P}^0$, *there exists at most a denumerable number of hyperplanes in* $\mathbf{R}_+^N$ *outside of which any point* x *is stable.*

Proof For all ergodic $\wedge$, we shall construct a system of hyperplanes (in fact restricted to $\mathbf{R}_+^N$) $B_1^\wedge, \dots B_{k(\wedge)}^\wedge$, $k(\wedge) \leq \infty$, satisfying the following properties:

(i) $B_i^\wedge$ belongs to the hyperplane generated by $B^\wedge$, but does not coincide with it, and $B_i^\wedge$ is parallel to $v^\wedge$, for $i = 1, 2, \dots, k(\wedge)$;

(ii) any $x \in B^{\wedge} - (\bigcup_i B_i^{\wedge})$ is stable but no other x is.

To construct such a system we proceed by induction from lower dimensions. First for any $\wedge, \wedge_1, \wedge_1$ being ergodic and such that $\wedge \subset \overline{\wedge}_1$, $\dim \wedge \leq \dim \wedge_1 - 2$, we consider the plane containing all segments $[x, x + \alpha v^{\wedge_1}], x \in \wedge_1$, if $\exists \alpha > 0$ such that $x + \alpha v^{\wedge_1} \in \wedge$.

If this plane does exist, then it has dimension $\dim \wedge + 1$ and contains all paths in $\wedge_1$ which enter $\wedge$ immediately after $\wedge_1$. All planes thus constructed will be called *first generation planes.*
We construct *second generation planes* in the following way: take an ergodic $\wedge_2 \neq \wedge_1$, $\dim \wedge_2 \geq \dim \wedge_1$, and, for any first generation plane in $\wedge_1$, we consider all paths in $\wedge_2$ which run along this plane immediately (possibly first intersecting some non-ergodic face) after $\wedge_2$. The plane of minimal dimension containing all such paths is a *second generation plane.* Similarly we construct *third generation planes,* etc. The planes of all generations comprise the desired system.
Now let $\mathcal{P}^0$ be such that, for any regular random walk in $\mathcal{P} - \mathcal{P}^0$, any $x \in B^{\wedge} - \bigcup_i B_i^{\wedge}$ (for an ergodic $\wedge$) is stable: the random walk belongs to $\mathcal{P}^0$ if, and only if, there exist $\wedge$ and $B_i^{\wedge}$ containing some subface of the face $B^{\wedge}$. For example, in $\mathbf{R}_+^3$ this could be a plane passing through two one-dimensional planes (i.e. lines). It is clear that $\mathcal{P}_0$ is of Lebesgue measure zero. In $\mathbf{Z}_+^3$, $\mathcal{P}_0 = \emptyset$. The reader will find examples of non-empty $\mathcal{P}_0$ in $\mathbf{Z}^3$ (which can be viewed as the union of 8 octants $\mathbf{Z}_+^3$).
Note that, by stability, if $\wedge_i$ is ergodic then $\dim\wedge_{i+1} = \dim \wedge_i - 1$. Moreover if a random walk is in $\mathcal{P} - \mathcal{P}_0$, then the following possibilities exist for an ergodic $\wedge_i$:

(i) $\wedge_{i-1}$ is non-ergodic. Then $\wedge_{i-1}$ does not belong to any $B_j^{\wedge_i}$. So $\wedge_{i-2}$ does not belong to any $B_j^{\wedge_{i-2}}$;

(ii) $\wedge_{i-1}$ is ergodic. Then $\wedge_{i-1}$ does not belong to any $B_j^{\wedge_{i-1}}$.

Hence, properties (i) and (ii) follow by construction and the proposition is proved. ∎

Unless otherwise stated, *we consider only strongly regular random walks,* but many definitions will be appropriate for more general situation. Some random walks in dimensions 3 and 4 are not strongly regular (see chapter 5) but proposition 4.2.3 can be easily reformulated for such cases.

Problem For some nonregular random walks, similar results could be proved. Describe the situation for an arbitrary random walk.

Remark The above construction vaguely resembles the construction of billiard dynamical systems [KSF80].

Definition 4.2.4 *A random walk is called* deterministic *if its second vector field has no branch points. A regular random walk is called* essentially deterministic *(e.d.r.w) if, for any stable x, Γ_x has no branch points.*

Let us consider an e.d.r.w. in $\mathcal{P} - \mathcal{P}_0$ and let T_t be the dynamical system on $\mathbf{R}_+^N$ defined by

$$T_t x = \Gamma_x(t) ,$$

for stable x, i.e. a point moves with velocity $v^\wedge$ along the ergodic face $\wedge$. For non stable x we shall define $T_t x$ by continuity at every point wherever possible and in any suitable way otherwise. Let us first note the following crucial scaling property: for any $\alpha > 0, t > 0$,

$$\Gamma_x(t) = \alpha^{-1}\Gamma_{\alpha x}(\alpha t) , \tag{4.4}$$

we can assume that T_t is defined so that (4.4) holds not only for stable x but for all x, by making use of the function ϕ defined as in the preceding section.

4.3 Gluing Lyapounov functions together

Let $Q = \{\xi_n\}$ be an irreducible aperiodic ergodic Markov chain, defined on some probability space $(\Omega_1, \Sigma_1, \mu_1)$, with countable state space $\mathcal{A}, \pi(\alpha)$ being its stationary probabilities.
Let us consider, on some other probability space $(\Omega_2, \Sigma_2, \mu_2)$, mutually independent vector-valued (with values in $\mathbf{R}^k$) random variables $g_{t,\alpha}(\omega_2), \alpha \in \mathcal{A}$, $t = 0, 1, \ldots$, and $\omega_2 \in \Omega_2$, indexed by t, α. The distribution of $g_{t,\alpha}$ does not depend on t, so that we can write $E[g_{t,\alpha}] \equiv F(\alpha)$. We assume $F(\alpha)$ is finite and we put

$$X_T = c + \sum_{t=0}^{T-1} g_{t,\xi_t} , \quad c \in \mathbf{R}^k ,$$

on $\{\Omega_1 \times \Omega_2, \Sigma_1 \times \Sigma_2, \mu_1 \times \mu_2\}$. As g_{t,ξ_t} is a stationary process, Birkhoff's ergodic theorem yields

$$\frac{X_T}{T} \to \sum_{\alpha \in \mathcal{A}} \pi(\alpha) F(\alpha) , \text{ a.s. when } T \to \infty .$$

The next result also follows from Birkhoff's ergodic theorem.

Lemma 4.3.1 *Assume the chain Q is ergodic and $\xi_0 = c$ at time $t = 0$. Then, for any $\epsilon > 0$,*

$$P\left(\| \sum_{t=0}^{n} g_{t,\xi_t} - n \sum_{\alpha \in \mathcal{A}} \pi(\alpha) F(\alpha) \| > n\epsilon \right) \to 0\,, \quad as \quad n \to \infty\,. \tag{4.5}$$

The proof is immediate using a convergence in probability argument. ■
Let us put for any $\wedge$, $c > 0$, $t > 0$,

$$B_{ct}^{\wedge} = \{(r_1, \ldots, r_N) : r_i > c,\ i \in \wedge; r_i \leq t,\ i \notin \wedge\}.$$

By means of lemma 4.3.1, we can prove the following assertion concerning random walks. Let ξ_m be the location of the random walk $\mathcal{L}$ in the m-th step with $\xi^0 = \alpha$.

Lemma 4.3.2 *Assume that either $\mathcal{L}^{\wedge}$ is ergodic for $| \wedge | < N$ or $\wedge = \{1, 2, \ldots, N\}$. Then, for any R_2 and $\epsilon, \sigma > 0$ there exist a natural number m and $R_1 > 0$ such that, for any point $\alpha \in B_{R_1 R_2}^{\wedge} \cap \mathbf{Z}_+^N$,*

$$P\Big(\| \xi_m - (\alpha + mv^{\wedge}) \| > m\epsilon \, | \xi_0 = \alpha \Big) < \sigma\,. \tag{4.6}$$

Proof In order to apply lemma 4.3.1, consider the chain $Q^{\wedge}$ whose states are all the one-step transitions of $\mathcal{L}^{\wedge}$, i.e. the pairs (a_i, a_j) for $a_i, a_j \in C^{\wedge}$. The transition probabilities of $Q^{\wedge}$ are defined as

$$P_{(a_1 a_2)(a_3 a_4)} = \begin{cases} 0, & a_2 \neq a_3\,, \\ {}_{\wedge}p_{a_2, a_4}, & a_2 = a_3\,. \end{cases}$$

Let $\pi_a (a \in C^{\wedge})$ be the stationary probabilities of $\mathcal{L}^{\wedge}$ and $\pi_{(a,b)}$, $(a, b) \in C^{\wedge} \times C^{\wedge}$, the stationary probabilities of $Q^{\wedge}$. It is obvious that

$$\pi_{(a,b)} = {}_{\wedge} p_{ab} \pi_a\,. \tag{4.7}$$

Let $\xi_0 = \alpha, \xi_1, \xi_2, \ldots$ be the sequence of random variables corresponding to the random walk $\mathcal{L}$, ξ_m^i being the i-th component of ξ_m, and let $\wedge = (i_1, \ldots, i_k)$. We introduce the sequence of random variables g_{m,ξ_m} by setting

$$g_{m,\xi_m} = (\xi_{m+1}^{i_1} - \xi_m^{i_1}, \xi_{m+1}^{i_2} - \xi_m^{i_2}, \ldots, \xi_{m+1}^{i_k} - \xi_m^{i_k})\,.$$

This random sequences satisfies the hypotheses of lemma 4.3.1. Using it, we obtain the assertion of lemma 4.3.2. ■

On the set $\mathbf{Z}_+^N$, let a real function $f(\alpha)$, $\alpha \in \mathbf{Z}_+^N$, be given such that the condition $|f_\alpha - f_\beta| > d$ implies that $p_{\alpha\beta} = 0$, for some $d > 0$. Introduce

$$B_R^{\wedge} = \{(x_1, \ldots, x_N) : x_i > R,\ i \in \wedge\}\,.$$

Lemma 4.3.3 *Assume that the chain $\mathcal{L}^\wedge$ is not ergodic, and there exist a set $B^\wedge_{R_1R_2}$ and a function $m(\alpha)$ defined on the set $(B^\wedge_{R_1} \setminus B^\wedge_{R_1R_2}) \cap \mathbf{Z}^N_+$ and taking values in the set of natural numbers such that, for all $\alpha \in (B^\wedge_{R_1} \setminus B^\wedge_{R_1R_2}) \cap \mathbf{Z}^N_+$, the inequality*

$$\sum_{\beta \in \mathbf{Z}^N_+} p^{m(\alpha)}_{\alpha\beta} f_\beta - f_\alpha < -\epsilon \tag{4.8}$$

holds for some $\epsilon > 0$, and

$$\sup_{\alpha \in (B^\wedge_{R_1} \setminus B^\wedge_{R_1R_2}) \cap Z^N_+} m(\alpha) = m < \infty\,. \tag{4.9}$$

Then there exist a set $B^\wedge_R$ and a function $n(\alpha)$, $\alpha \in B^\wedge_R$, taking values in the set of natural numbers, such that, for all $\alpha \in B^\wedge_R \cap \mathbf{Z}^N_+$, the inequality

$$\sum_{\beta \in Z^N_+} p^{n(\alpha)}_{\alpha\beta} f_\beta - f_\alpha < -\epsilon_1 \tag{4.10}$$

holds for some $\epsilon_1 > 0$, and

$$\sup_{\alpha \in B^\wedge_R \cap Z^N_+} n(\alpha) \equiv n < \infty;\ n(\alpha) \equiv m(\alpha),\ \alpha \in B^\wedge_R \setminus B^\wedge_{R_1R_2}\,.$$

Proof Let $\xi_0 = \alpha\,, \xi_1\,, \xi_2, \ldots$ be the sequence of random variables corresponding to the chain $\mathcal{L}$. Form the random index sequence N_i by setting $N_0 = m(\alpha)$ and $N_i = N_{i-1} + m(\xi_{i-1})$ (for $\alpha \in \mathbf{Z}^N_+ \setminus B^\wedge_{R_1R_2}$, we complete the definition of $m(\alpha)$ by setting $m(\alpha) = 1$). The sequence ξ_{N_i} forms a Markov chain $\tilde{\mathcal{L}}$. Accordingly $\tilde{\mathcal{L}}^\wedge$ denotes the Markov chain induced by $\tilde{\mathcal{L}}$ on the state set $C^\wedge$. It is obvious that, if $\{\xi^\wedge_i\}$ is the sequence of random variables corresponding to $\mathcal{L}^\wedge$, then the sequence $\{\xi^\wedge_{N_i}\}$ corresponds to $\tilde{\mathcal{L}}^\wedge$. The non-ergodicity of $\mathcal{L}^\wedge$ implies that of $\tilde{\mathcal{L}}^\wedge$. Therefore, for any $\sigma > 0$, there exist $R \gg R_1$ and $t > 0$ such that, for all r with $\dfrac{R - R_1}{m} > r > t$, we have

$$P(\tilde{\xi}_r \notin B^\wedge_{R_1R_2}) > 1 - \sigma\,, \tag{4.11}$$

provided that $\xi_0 = \alpha \in B^\wedge_{RR_2}$. Taking account of (4.8), we obtain from (4.11) that

$$\begin{aligned} &E\left(f(\tilde{\xi}_r) - f(\tilde{\xi}_{r-1})\right) \\ &\quad = E\left(f(\tilde{\xi}_r) - f(\tilde{\xi}_{r-1}) | \tilde{\xi}_{r-1} \notin B^\wedge_{R_1R_2}\right) P(\tilde{\xi}_{r-1} \notin B^\wedge_{R_1R_2}) \\ &\quad + E\left(f(\tilde{\xi}_r) - f(\tilde{\xi}_{r-1}) | \tilde{\xi}_{r-1} \in B^\wedge_{R_1R_2}\right) P(\tilde{\xi}_{r-1} \in B^\wedge_{R_1R_2}) \\ &\quad \le dm\sigma - \epsilon(1 - \sigma)\,. \end{aligned}$$

Consequently, choosing σ sufficiently small (hence R must be taken sufficiently large), we obtain

$$E(f(\tilde{\xi}_r) - f(\tilde{\xi}_{r-1})) < -\sigma_1 ,$$

for some $\sigma_1 > 0$ and $(R - R_1)/m > r > t$; $\xi_0 = \alpha \in B^{\wedge}_{RR_2}$. Moreover,

$$E(f(\tilde{\xi}_r)) - f_\alpha = \sum_{i=1}^{r} E(f(\tilde{\xi}_i) - f(\tilde{\xi}_{i-1})) < tdm - \sigma_1(r - t) .$$

Thus, for any $\alpha \in B^{\wedge}_{RR_2}$, there exists $\tilde{m}(\alpha) \in \mathbf{Z}_+$ such that

$$E\left(f(\tilde{\xi}_{\tilde{m}(\alpha)})|\tilde{\xi}_0 = \alpha\right) \leq f(\alpha) - \tilde{\epsilon} , \tag{4.12}$$

where $\tilde{\epsilon} > 0$ does not depend on α.
We use (4.12) and the exponential estimates of theorem 2.1.8. Choosing $\sigma > 0$, it follows easily that there exists $\delta > 0, t \in \mathbf{Z}_+$ and $R > 0$ such that, for any $\alpha \in B^{\wedge}_{RR_2}$,

$$P\left(\sup_{t \leq r \leq \frac{R-R_1}{m}} (f(\tilde{\xi}_r) + \delta r) \leq f(\alpha) / \xi_0 = \alpha\right) > 1 - \sigma . \tag{4.13}$$

Note now that

$$f(\xi_k) \leq \sup_{k/m < r < k} f(\tilde{\xi}_r) + md . \tag{4.14}$$

From (4.13) and (4.14), for sufficiently large R and k, one can easily get

$$E(f(\xi_k)/\xi_0 = \alpha) \leq f(\alpha) - \epsilon_1 , \tag{4.15}$$

where $\epsilon_1 > 0$ does not depend on $\alpha \in B^{\wedge}_{RR_2}$.
Inequality (4.15) is equivalent to (4.10) if we put $n(\alpha) = k$, for $\alpha \in B^{\wedge}_{RR_2} \cap \mathbf{Z}^N_+$, in (4.15). For $\alpha \in (B^{\wedge}_R \setminus B^{\wedge}_{RR_2}) \cap \mathbf{Z}^N_+$, we set $n(\alpha) \equiv m(\alpha)$. The lemma is proved. ■

We have previously introduced a finite collection $\{v^\wedge\}$ of vectors. To each point α belonging to an ergodic face $B^\wedge$, we assign the vector $v(\alpha) = v^\wedge$. For the points $\alpha \in B^{\{1,2,\ldots,N\}}$, we set $v(\alpha) = v^{\{1,2,\ldots,N\}}$. In this way, we obtain a vector field V, which may not be defined on certain faces.

Condition B: For some $\delta, b, p > 0$, there exists a function $f(\alpha)$, $\alpha \in \mathbf{R}^N_+$ having the following properties:

(i) $f(\alpha) \geq 0, \quad \alpha \in \mathbf{R}^N_+$;
(ii) $f(\alpha) - f(\beta) \leq b \parallel \alpha - \beta \parallel, \quad \alpha, \beta \in \mathbf{R}^N_+$;

(iii) for any $\wedge$ such that $\mathcal{L}^\wedge$ is ergodic (including $\wedge = \{1, \ldots, N\}$) and all $\alpha \in B_{pp}^\wedge$,

$$f(\alpha + v(\alpha)) - f(\alpha) < -\delta \, .$$

Condition B′: For some $\delta, b, t, p > 0$ there exist a function $f(\alpha)$, $\alpha \in \mathbf{R}_+^N$, and a nonempty set $T \subset \mathbf{R}_+^N$, satisfying the following conditions:

(i) $f(\alpha) \geq 0, \quad \alpha \in \mathbf{R}_+^N$;

(ii) $f(\alpha) - f(\beta) \leq b \parallel \alpha - \beta \parallel, \quad \alpha, \beta \in \mathbf{R}_+^N$;

(iii) $f(\alpha) \geq t, \quad \alpha \in T$;
$f(\alpha) < t, \quad \alpha \in R_+^N \backslash T$;

(iv) for any $\wedge$ such that $\mathcal{L}^\wedge$ is ergodic, for $\wedge = \{1, \ldots, N\}$ and all $\alpha \in \bigcap B_{pp}^\wedge \cap T$, we have

$$f(\alpha + v(\alpha)) - f(\alpha) > \delta \, .$$

Theorem 4.3.4 *If the vector field V satisfies condition* **B**, *then the random walk $\mathcal{L}$ is ergodic. If condition* **B′** *is satisfied, then $\mathcal{L}$ is transient.*

Proof Assume there exists a function $f(\alpha)$, $\alpha \in \mathbf{R}_+^N$, satisfying condition **B** . As follows from corollary 2.2.5, for the ergodicity of $\mathcal{L}$ it is sufficient to show the existence of a function $m(\alpha)$, $(\alpha \in \mathbf{Z}_+^N)$, taking values in the set of natural numbers, such that

$$\sup_{\alpha \in \mathbf{Z}_+^N} m(\alpha) = m < \infty$$

and, for all $\alpha \in \mathbf{Z}_+^N$ except some finite set, the inequality

$$\sum_{\beta \in \mathbf{Z}_+^N} p_{\alpha\beta}^{m(\alpha)} f_\beta - f_\alpha < -\epsilon_1 \qquad (4.16)$$

is satisfied for some $\epsilon_1 > 0$. Let $\wedge = \{1, 2, \ldots, N\}$. It follows from lemma 4.3.2 that for any $\epsilon, \sigma > 0$ there exist $m^\wedge$ and $R^\wedge$ such that inequality (4.6) is satisfied for all $\alpha \in B_{R^\wedge}^\wedge$. Therefore, choosing ϵ and σ sufficiently small, taking into account the boundedness of the jumps of the random walk and the properties of the function f (condition **B**), we obtain that, for any $\alpha \in B_{R^\wedge}^\wedge \cap \mathbf{Z}_+^N$, inequality (4.16) is satisfied for some $\epsilon_1 > 0$ by setting

$$m(\alpha) \equiv m^\wedge \, , \quad \text{for } \alpha \in B_{R^\wedge}^\wedge \cap \mathbf{Z}_+^N \, .$$

We continue the construction of $m(\alpha)$, $(\alpha \in \mathbf{Z}_+^N)$ by induction. Assume that for all $\wedge, |\wedge| = k \leq N$, there exist sets $B_{R^\wedge}^\wedge$ and a function $m(\alpha)$ with values in the set of natural numbers such that

$$\sup_{\alpha \in \bigcup_{|\wedge|=k} B_{R^\wedge}^\wedge \cap \mathbf{Z}_+^N} m(\alpha) < \infty$$

and, for all $\alpha \in \bigcup_{|\wedge|=k} B_{R^\wedge}^\wedge \cap \mathbf{Z}_+^N$, inequality (4.16) is satisfied for some $\epsilon_1 > 0$. Take $\wedge_1$ such that $|\wedge_1| = k-1$. It follows from the definition of the sets $B_{R_1 R_2}^\wedge$ and $B_R^\wedge$ that there exist $R_1^{\wedge_1}, R_2^{\wedge_1} > 0$, and together with them sets $B_{R_1^{\wedge_1} R_2^{\wedge_1}}^{\wedge_1}$, such that

$$B_{R_1^{\wedge_1}}^{\wedge_1} \backslash B_{R_1^{\wedge_1} R_2^{\wedge_1}}^{\wedge_1} \subset \bigcup_{|\wedge|=k} B_{R^\wedge}^\wedge . \tag{4.17}$$

(a) Let $\mathcal{L}^{\wedge_1}$ be non-ergodic. Then, applying lemma 4.3.3 and using (4.17), we conclude that there exist $R^{\wedge_1}$ and a function $n(\alpha)$ with values in the set of natural numbers such that

$$\sup_{\alpha \in Z_+^N} n(\alpha) \quad < \quad \infty ,$$

$$n(\alpha) \equiv m(\alpha) \quad , \quad \text{for } \alpha \in \left(B_{R^{\wedge_1}}^{\wedge_1} \backslash B_{R_1^{\wedge_1} R_2^{\wedge_1}}^{\wedge_1}\right) \cap \mathbf{Z}_+^N ,$$

and

$$\sum_{\beta \in Z_+^N} p_{\alpha\beta}^{n(\alpha)} f_\beta - f_\alpha < -\epsilon^{\wedge_1} , \tag{4.18}$$

for all $\alpha \in B_{R^{\wedge_1}}^{\wedge_1}$ and some $\epsilon_1^{\wedge_1} > 0$.

(b) Let $\mathcal{L}^{\wedge_1}$ be ergodic. In this case we use lemma 4.3.2 and we obtain the same result as in the case where $\mathcal{L}^{\wedge_1}$ is non-ergodic.

Sorting out all $\wedge$ with $|\wedge| = k-1$, we obtain that the conditions which we assumed to be satisfied for all $\wedge$ with $|\wedge| = k$ will also be satisfied for all $\wedge$ with $|\wedge| = k-1$.
Hence, we have proved by induction that, for all $\wedge$ with $|\wedge| = 1$, there exist sets $B_{R^\wedge}^\wedge$, for some $R^\wedge > 0$, and a function $m(\alpha)$, taking its values in the set of natural numbers, such that

$$\sup_{\alpha \in \bigcup_{|\wedge|=1} B_{R^\wedge}^\wedge \cap Z_+^N} m(\alpha) < \infty ,$$

and for all $\alpha \subset \bigcup_{|\wedge|=1} B_{R^\wedge}^\wedge$ inequality (4.16) is satisfied for some $\epsilon_1 > 0$. This completes the proof of the ergodicity of the random walk, if we take into account that $\mathbf{Z}_+^N \setminus \bigcup_{|\wedge|=1} B_{R^\wedge}^\wedge$ is a finite set.

Assume now that there exists a function $f(\alpha)$, $\alpha \in \mathbf{R}_+^N$, satisfying condition $\mathbf{B}'$. From theorem 2.2.7, it suffices to prove the transience of $\mathcal{L}$ to show the existence of a function $m(\alpha)$, $\alpha \in \mathbf{Z}_+^N$, with values in the set of natural numbers, such that

$$\sup_{\alpha \in \mathbf{Z}_+^N} m(\alpha) = m < \infty$$

and for all $\alpha \in T$, except some finite set, the inequality

$$\sum_{\beta \in \mathbf{Z}_+^N} p_{\alpha\beta}^{m(\alpha)} f_\beta - f_\alpha > \epsilon_1 \tag{4.19}$$

is satisfied, for some $\epsilon_1 > 0$. The proof of the existence of $m(\alpha)$ can be carried out by induction analogously and in the same succession as in the ergodic case. The theorem is proved. ■

Remark The results obtained in the above sections are in fact valid under the following more general assumptions:

Partial homogeneity. Condition A_0 of section 3.2 can be replaced by the following one: There exists $c > 0$ such that, for any Λ and for all $a \in B^\Lambda \cap \mathbf{Z}_+^N$,

$$p_{\alpha\beta} = p_{\alpha+a,\beta+a}, \quad \forall \alpha \in B_{cc}^\Lambda \cap \mathbf{Z}_+^N, \ \forall \beta \in Z_+^N.$$

Lower boundedness of the jump + first moment condition. Condition A_1 of section 3.2 can be replaced by the two conditions

$$p_{\alpha\beta} = 0, \text{ for } (\alpha_i - \beta_i) > -d, \ \forall i = 1, \ldots, N,$$

$$\sum_\beta (\beta - \alpha) p_{\alpha\beta} < \infty, \forall \beta \in \mathbf{Z}_+^N,$$

where d is some positive constant.

4.4 Classification in $\mathbf{Z}_+^3$

Before considering dimension 3 let us come back to random walks in $\mathbf{Z}_+^2$ and look at them from the point of view of induced chains. In fig. 4.4.1, some possible directions of the vectors v^Λ are shown. Note that random walks under conditions $\mathbf{0}_1$ and $\mathbf{0}_2$ are always *deterministic.*

(i) Case 1: $v^{\{1,2\}}$ has both components positive. So neither one-dimensional face is ergodic.

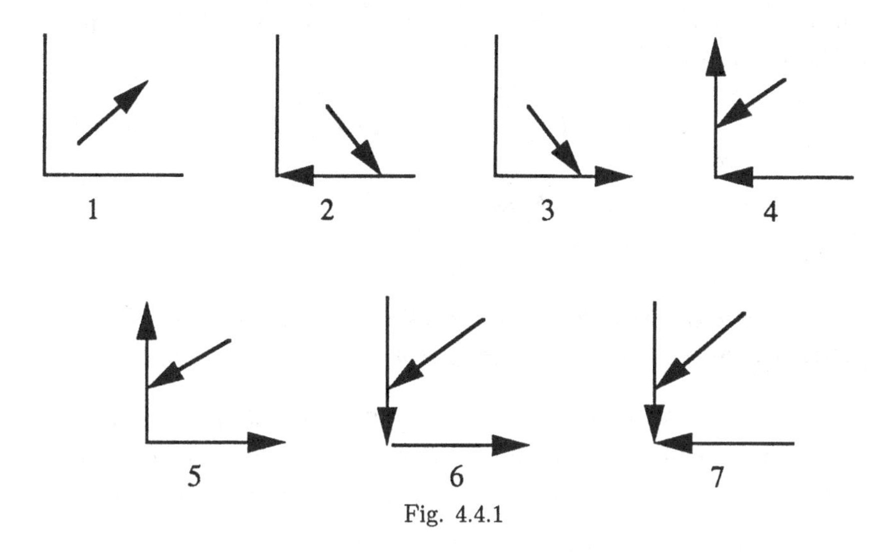

Fig. 4.4.1

(ii) Cases 2–3: One component ($i = 1$) of $v^{\{1,2\}}$ is positive, the other one ($i = 2$) is negative. So the face $\wedge(1)$ is ergodic, $\wedge(2)$ is not.

(iii) Cases 4–7: Both one-dimensional faces are ergodic.

The main result of section 3.3 can be reformulated in the following way.

Theorem 4.4.1 *Under assumptions* $\mathbf{0}_1$ *and* $\mathbf{0}_2$*, the random walk in* $\mathbf{Z}_+^2$ *is transient if, and only if, there exists a path* Γ_x *going to infinity. Otherwise (when all paths go to 0) it is ergodic.*

Proof If there is a path going to infinity then (see fig. 4.4.1), there exists a face $\wedge$ (one- or two-dimensional) such that $v^\wedge$ has all its components positive, so that the random walk is transient by proposition 4.1.4. On fig. 4.4.1 cases 1, 3–6 are transient and we are left with the cases 2 and 7.
In these cases, for each point x, the first passage time $\tau(x)$ of the dynamical system $\Gamma(t) \equiv \Gamma_x(t)$ to the origin is finite. We introduce the function $f(x)$, on $\mathbf{R}_+^2$, by putting $f(x) = \tau(x)$. It is then easy to see that $f(x)$ satisfies condition $\mathbf{B}$ of the preceding section. Consequently, $\mathcal{L}$ is ergodic and the theorem is proved. ∎

The following theorem 4.4.2 is just a rephrasing of theorem 4.4.1 in terms of the second vector field.

Theorem 4.4.2 *The random walk $\mathcal{L}$ is ergodic if, and only if, the following two conditions are satisfied:*

(i) *there exists an i_1 such that $v_{i_1}^{\{1,2\}} < 0$;*

(ii) $v_{i_2}^{\{i_1\}} < 0$ *for very i_1 such that $v_{i_1}^{\{1,2\}} < 0$.*

Let us pass to the study of random walks in $\mathbf{Z}_+^3$. In this case, the second vector field also has a simple construction.

Lemma 4.4.3 *For every point $x \in \mathbf{R}_+^3 \backslash 0$ the vector field V is defined. Moreover, single-valuedness can be violated only on one of the three one-dimensional faces $B^{\{i_1\}}$. In this case only two vectors can be assigned to the points of $B^{\{i_1\}}$.*

Proof Let $x \in B^{\{1,2,3\}}$. Then $v(x) = M(\alpha) = M^{\{1,2,3\}}$, where $M^{\{1,2,3\}}$ is the mean drift from the point $\alpha \in B^{\{1,2,3\}} \cap \mathbf{Z}_+^3$.
For $x \in B^{\{i_1,i_2\}}$, we put $v(x) = M^{\{1,2,3\}} = (M_1^{\{1,2,3\}}, M_2^{\{1,2,3\}}, M_3^{\{1,2,3\}})$ if $M_{i_3}^{\{1,2,3\}} > 0$. In this case the induced chain $\mathcal{L}^{\{i_1,i_2\}}$ is transient. When $M_{i_3}^{\{1,2,3\}} < 0$, the induced chain $\mathcal{L}^{\{i_1,i_2\}}$ is ergodic and so, for $x \in B^{\{i_1,i_2\}}$, the vector $v(x)$ is uniquely defined. For the points which belong to one-dimensional faces the situation is different. If $\mathcal{L}^{\{i_1\}}$ is ergodic, then the vector $v(x)$ for $x \in B^{\{i_1\}}$ is uniquely defined. If $\mathcal{L}^{\{i_1\}}$ is transient, then the uniqueness is violated only when the faces $B^{\{i_1,i_2\}}$ and $B^{\{i_1,i_3\}}$ are ergodic and $M_{i_2}^{\{i_1,i_2\}} > 0$, $M_{i_3}^{\{i_1,i_3\}} > 0$. But this situation can happen only on a single one-dimensional face. The lemma is proved ∎

If the vector field V is single-valued, then it gives rise, in a natural way, to a dynamical system in $\mathbf{R}_+^3$ for which V is the velocity field. On the other hand, if the single-valuedness of V is violated on a one-dimensional face $B^{\{i_1\}}$, then we choose one of the vectors assigned to $B^{\{i_1\}}$. In this way we obtain two single-valued fields V_1 and V_2. For each field V_i, we construct a dynamical system $\Gamma^i(t)$.

Theorem 4.4.4 *If for any point $x \in \mathbf{R}_+^3$ the first passage time $\tau(x)$ for the dynamical system $\Gamma^i(t)$, $\Gamma^i(0) = x$, to reach the origin is finite for at least one $i, i = 1, 2$, then $\mathcal{L}$ is ergodic. On the other hand, if at least one dynamical system is such that $\tau(x) = \infty$, then $\mathcal{L}$ is transient.*

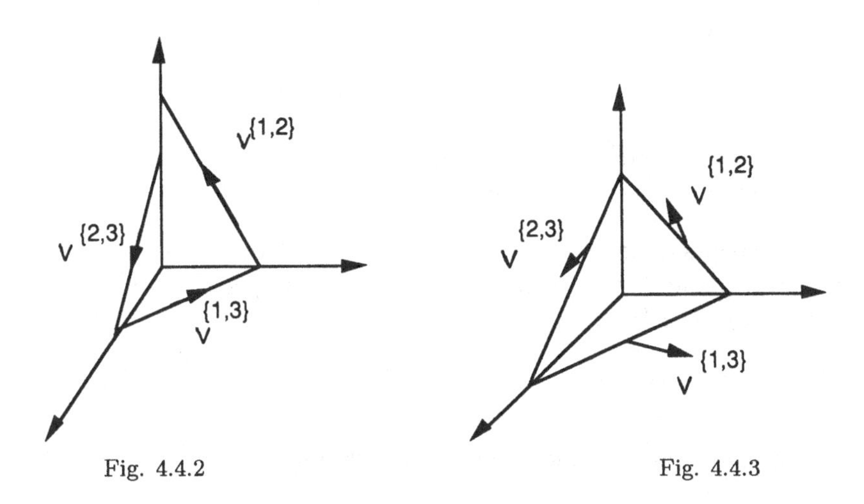

Fig. 4.4.2 Fig. 4.4.3

Proof Let $\tau_i(x)$ be finite for any point $x \in \mathbf{R}_+^3$. As in theorem 4.4.1 for $\mathbf{R}_+^2$, we introduce a function $f(x)$ by setting $f(x) = \tau_i(x)$. If V is single-valued, then this function satisfies condition **B** of section 4.3 and so $\mathcal{L}$ is ergodic.

If the single-valuedness of V is violated on a one-dimensional face $B^{\{i_1\}}$, then the continuity of $f(x)$ is violated on the plane going through the face $B^{\{i_1\}}$ and the vector $v^{\{1,2,3\}}$. However, this discontinuity is easily removable. For this, the values of $f(x)$ have to be multiplied by the corresponding factor on one side of the indicated plane. The function thus corrected, again satisfies condition **B**, which leads to the ergodicity of $\mathcal{L}$.

Let now $\tau_i(x)$ be infinite for some $x \in \mathbf{R}_+^3$. If x belongs to $B^{\{1,2,3\}}$ then transience follows from theorem 4.4.1. If $\tau_i(x)$ is infinite due to the fact that, for some ergodic face $B^\wedge$, all components of $V^\wedge$ are positive, then the random walk is transient for the same reason. It remains to consider the last possibility making $\tau_i(x) = \infty$. It is the case when the dynamical system, starting from some point, reaches one of the two-dimensional faces in a finite time and continues successively intersecting all two-dimensional and one-dimensional faces, in tending to infinity. This case is shown in fig. 4.4.2, and now we will construct the function

$f(x), x \in \mathbf{R}_+^3$, which satisfies the condition $\mathbf{B}'$ and then we use theorem 4.3.4. In the case shown in fig.4.4.2, all components of the vector $V^{\{1,2,3\}} = M^{\{1,2,3\}}$ are negative. In fig. 4.4.3, we show the surface $\{x : f(x) = a\}$.
Consider a triangle 123 with the following properties:

(i) its vertices 1,2,3 belong to corresponging one-dimensional faces, $i \in \wedge(i)$;
(ii) the second vector field on two-dimensional faces points to the outside (toward ∞) of this triangle.

Such triangles exist. Then we choose one and consider one of its sides, e.g. 12, belonging to the face $\wedge(1,2)$. Consider a plane H_{12} with the following properties:

(i) H_{12} contains the side 12;
(ii) H_{12} is sufficiently close to $\wedge(1,2)$;
(iii) $H_{12} \cap \mathbf{R}_+^3$ has an infinite Lebesgue measure.

Two other planes H_{23} and H_{13} can be defined in the same way. Now we define the level surface

$$H_1 = \{x : f(x) = 1\} = \bigcup_{(i,j)} \{H_{ij} \cap \mathbf{R}_+^3\} .$$

For any $a > 1$, put

$$H_a = \{x : f(x) = a\} = aH_1 .$$

Since we are interested in proving transience, we can simply put $f(x) = 1$ in the remaining part of $\mathbf{R}_+^3$. It is easy to see that the function $f(x)$ thus defined does satisfy condition $\mathbf{B}'$. The proof of theorem 4.4.4 is concluded. ∎

In fact the conditions of theorem 4.4.4 can be written in an explicit way.

Theorem 4.4.5 *The dynamical system $\Gamma^i(t)$ reaches the origin in a finite time for any initial state $\Gamma^i(0) = x \in \mathbf{R}_+^3$ if, and only if, the following three conditions are satisfied:*

(i) *There exists an i_1 such that $v_{i_1}^{\{1,2,3\}} < 0$.*
(ii) *For every i_1 such that $v_{i_1}^{\{1,2,3\}} < 0$, there exists an i_2 for which $v_{i_2}^{\{i_2,i_3\}} < 0$.*

(iii) *Either*

(a) there exists an i_1 such that $L^{\{i_1\}}$ is ergodic and, for any i_1 such that $L^{\{i_1\}}$ is ergodic, we have $v_{i_1}^{\{i_1\}} < 0$; or

(b) the chains $L^{\{1\}}, L^{\{2\}}, L^{\{3\}}$ are transient; if $v_2^{\{2,3\}} > 0$, then

$$\left| \frac{v_2^{\{2,3\}}}{v_3^{\{2,3\}}} \frac{v_3^{\{1,3\}}}{v_1^{\{1,3\}}} \frac{v_1^{\{1,2\}}}{v_2^{\{1,2\}}} \right| < 1 ,$$

and, if $v_2^{\{2,3\}} < 0$, then

$$\left| \frac{v_2^{\{2,3\}}}{v_3^{\{2,3\}}} \frac{v_3^{\{1,3\}}}{v_1^{\{1,3\}}} \frac{v_1^{\{1,2\}}}{v_2^{\{1,2\}}} \right| > 1 .$$

(The case where the expression between the absolute value signs is equal to 1 is not considered.)

Remark It follows that, if there exist two dynamical systems and if one of them has the above property, then so does the other one.

5
Random walks in two-dimensional complexes

5.1 Introduction and preliminary results

A two-dimensional complex is a union of a finite number of quarter-planes $\mathbf{Z}_+^2$ having some boundaries in common. An example can be the union of all two-dimensional faces of $\mathbf{Z}_+^N$. We consider maximally homogeneous random walks on such complexes and obtain necessary and sufficient conditions for ergodicity, null recurrence and transience up to some *non-zero* assumptions which are of measure 1 in the parameter space.

In chapter 4, a vector field was constructed allowing for a complete classification to be obtained for $N = 2, 3$. Moreover sufficient conditions for ergodicity and transience were derived for $N > 3$. One of the main features of the vector field in question is that it was deterministic. In this chapter, certain vector fields appear, which are deterministic inside two-dimensional faces, but give rise to random scattering on one-dimensional faces. We shall see that the calculation of the exit boundary of some countable one-dimensional Markov chain is necessary because of this phenomenon. This is the first new phenomenon, which is common also for $\mathbf{Z}_+^N$. The second new phenomenon is that null recurrence exists for a set of parameters which has a positive measure in the parameter space. We give an explicit solution to our problem. In fact it reduces to finding stationary probabilities for a finite Markov chain with n states, where n is the number of two-dimensional faces in the simplex considered, and to calculation of the maximal eigenvalue of some $n \times n$ matrix with positive entries.

The chapter is organized as follows:

After the main definitions and preliminary results in sections 5.1 and 5.2, we formulate the key theorems 5.3.2 and 5.3.4. The proof of the ergodicity conditions, using a method of pasting local Lyapounov functions as explained in chapter 3, is given in section 5.5. Transience is proved in section 5.6, by using a simpler method to construct more global Lyapounov functions. Proofs of recurrence and non-ergodicity are presented in sections 5.7 and 5.8. An example of interacting queues, submitted to different regimes, is worked out in detail in section 5.9. Possible generalizations are briefly described in section 5.10.

We give now the essential definitions and quote some preliminary results.

Definition 5.1.1 *We call a* two-dimensional complex $\tilde{T}$ *any union of finite number of copies of* $\mathbf{R}_+^2$*:*

$$(\mathbf{R}_+^2)_i = \{(x_1, x_2)_i : x_j \geq 0\} \ , \ i = 1, \ldots, n \, .$$

We assume that all origins are identified, i.e. $(0,0)_i \stackrel{\text{def}}{=} \underline{0}$ for all i. Also some pairs $(\tilde{\Lambda}_{ji}^{(1)}, \tilde{\Lambda}_{j_1 i_1}^{(1)})$ of one-dimensional faces of $\tilde{T}$,

$$\tilde{\Lambda}_{1i}^{(1)} = \{(x_1, 0)_i : x_1 > 0\}, \ \tilde{\Lambda}_{2i}^{(1)} = \{(0, x_2)_i : x_2 > 0\},$$

can be identified as well. This means precisely that the points of $\tilde{\Lambda}_{ji}^{(1)}$ and $\tilde{\Lambda}_{j_1 i_1}^{(1)}$ lying at the same distance from the origin are identified. We shall consider discrete time homogeneous Markov chains $\mathcal{L} = \mathcal{L}_T$ with state space T, the integer points of $\tilde{T}$, i.e. the union of all

$$(\mathbf{Z}_+^2)_i = \{(x_i, x_2)_i : x_j \geq 0 \text{ integers } \} \subset (\mathbf{R}_+^2)_i \ ,$$

taking into consideration the above identifications between them. We denote the interior part of a generic two-dimensional face by $\tilde{\Lambda}^{(2)}$, e.g

$$\tilde{\Lambda}_1^{(2)} = \{(x_1, x_2)_1 : x_1, \, x_2 > 0\} \, .$$

Introduce also $\overline{\Lambda}^{(1)}, \overline{\Lambda}^{(2)}$ — the closures of $\tilde{\Lambda}^{(1)}, \tilde{\Lambda}^{(2)}$ — and $\Lambda^{(1)}, \Lambda^{(2)}$ – the sets of integer points of $\tilde{\Lambda}^{(1)}$ and $\tilde{\Lambda}^{(2)}$ respectively.

Examples

(i) $\tilde{T}$ (or T) is called *planar* if any one-dimensional face is identified with at most one other one-dimensional face. This means that $\tilde{T}$ can be topologically embedded into $\mathbf{R}^2$. Here are some examples:

  (a) $\mathbf{Z}_+^2$,

  (b) the union of four quadrants of $\mathbf{Z}^2$,

  (c) the union of three two-dimensional faces of $\mathbf{Z}_+^3$.

(ii) The union of all $\frac{N(N-1)}{2}$ two-dimensional faces of $\mathbf{Z}_+^N$, $N \geq 4$, is an example of a non-planar complex. The simplest example to have in mind is the union of five two-dimensional faces of $\mathbf{Z} \times \mathbf{Z}_+^2$, i.e. the union of two $\mathbf{Z}_+^3$ having a two-dimensional face in common.

(iii) $\tilde{T}$ is called strongly connected if it is not the union of two complexes $\tilde{T}_1$ and $\tilde{T}_2$ which have only $\underline{0}$ as a common point.

Let us note that in $\tilde{T}$ any line has a length. Hence, $\tilde{T}$ is a metric space endowed with the distance $\rho(\alpha, \beta)$ between $\alpha, \beta \in \tilde{T}$, equal to the minimal length of a line between them. We assume that the one-step transition probabilites $p_{\alpha\beta}(\alpha \to \beta)$ (repeated here for the reader's convenience) satisfy the following conditions:

Condition A1 *(Boundedness of the jumps)*

(i) $p_{\alpha\beta} = 0$ if α, β do not belong to the same $\overline{\Lambda}^{(2)}$;
(ii) $p_{\alpha\beta} = 0$ if $\rho(\alpha, \beta) > d$ for some fixed $d < \infty$;
(iii) $p_{\alpha\beta} = 0$ if at least one component of the vector $\beta - \alpha$ is less than -1.

Condition A2 *(Maximal space homogeneity)*
Let α, α' belong to the same (open) face Λ which can be one- or two -dimensional. If $\Lambda \subset \overline{\Lambda}^{(2)}$ (i.e. $\Lambda = \Lambda^{(2)}$ or Λ is a one-dimensional face of $\Lambda^{(2)}$) and

$$\beta' - \alpha' = \beta - \alpha ,$$

then

$$p_{\alpha\beta} = p_{\alpha'\beta'} \stackrel{\text{def}}{=} p_{\beta-\alpha}^{\Lambda} .$$

Thus our Markov chain is uniquely specified by a finite number of parameters p_{γ}^{Λ}, with $\gamma \in \overline{\Lambda}^{(2)}$ such that $\Lambda \subset \overline{\Lambda}^{(2)}$. We shall also make some assumptions $\mathbf{0}_1, \ldots, \mathbf{0}_5$, which we call *non-zero assumptions* and which exclude from our consideration some hypersurfaces in the parameter space (in the sense that they are of measure zero in this parameter space). Some of these assumptions are made just for economy of space and time, but others are very essential and will appear now.

Assumption $\mathbf{0}_1$ The Markov chain $\mathcal{L}_T$ is supposed to be irreducible and aperiodic.

Then this chain is ergodic if, and only if, any of its strongly connected component is ergodic. Therefore, we shall consider only strongly connected complexes T. For any two-dimensional face Λ and any $\alpha \in \Lambda$, we define the vectors $M(\alpha)$, the one-step mean jumps from α. They are all equal to

$$M_\Lambda = \sum_{\beta \in \overline{\Lambda}} (\beta - \alpha) p^{\Lambda}_{\beta-\alpha} = M(\alpha), \ \forall \alpha \in \Lambda .$$

If $\alpha \in \Lambda^{(1)} \equiv \Lambda$, then we define $M_\Lambda = M(\alpha)$ to be the collection of vectors $M_{\Lambda,\Lambda}$ and $M_{\Lambda,\Lambda^{(2)}}$, for all $\overline{\Lambda^2} \supset \Lambda$, such that

$$\begin{aligned} M_{\Lambda,\Lambda} &= \sum_{\beta \in \overline{\Lambda}} (\beta - \alpha) p^{\Lambda}_{\beta-\alpha} , \\ M_{\Lambda,\Lambda^{(2)}} &= \sum_{\beta \in \Lambda^{(2)}} (\beta - \alpha) p^{\Lambda}_{\beta-\alpha} . \end{aligned}$$

If one can embed $\tilde{T}$ into $\mathbf{R}^N$ for some N, so that all $\Lambda^{(2)}$ are orthogonal, then $M(\alpha)$ for $\alpha \in \Lambda^{(1)}$ can be defined as the usual vector of mean jumps.

Theorem 5.1.2 *If, for at least one $\Lambda = \Lambda^{(2)}$, the vector M_Λ has both components positive then $\mathcal{L}_T$ is transient*

Assumption 0$_2$ For any Λ the vector M_Λ can have no zero component.

Definition 5.1.3 *Let $\Lambda^{(1)}$ be a one-dimensional face and $S(\Lambda^{(1)})$ be the set of all two-dimensional faces $\Lambda^{(2)}$ such that $\Lambda^{(1)} \subset \overline{\Lambda^{(2)}}$. Let $S_+(\Lambda^{(1)}) \subset S(\Lambda^{(1)})$ be the set of all $\Lambda^{(2)}$ such that $M_{\Lambda^{(2)}}$ looks onto $\Lambda^{(1)}$, i.e. its component perpendicular to $\Lambda^{(1)}$ is negative. Accordingly, $S_-(\Lambda^{(1)}) = S(\Lambda^{(1)}) - S_+(\Lambda^{(1)})$. We call $\Lambda^{(2)} \in S_+(\Lambda^{(1)})$ (resp. $S_-(\Lambda^{(1)})$) an* ingoing *(resp.* outgoing*) face for $\Lambda^{(1)}$. If $S_+(\Lambda^{(1)}) = S(\Lambda^{(1)})$, then $\Lambda^{(1)}$ is called ergodic.*

Definition 5.1.4 *Let us consider a one-dimensional face $\Lambda^{(1)}$ and a point $\alpha \in \Lambda^{(1)}$. For any two-dimensional $\Lambda \in S(\Lambda^{(1)})$, let us consider the half-line $C^{\Lambda}_{\Lambda^{(1)}}$ which belongs to $\overline{\Lambda}$ and is perpendicular to $\Lambda^{(1)}$ at the point α. We call a* hedgehog *the following one-dimensional complex:*

$$H_{\Lambda^{(1)}} = \bigcup_{\Lambda \in S(\Lambda^{(1)})} C^{\Lambda}_{\Lambda^{(1)}} .$$

For different $\alpha \in \Lambda^{(1)}$, these hedgehogs are congruent in the obvious sense. Let us consider the Markov chain $\mathcal{L}_{\Lambda^{(1)}}$, with set of states $H_{\Lambda^{(1)}}$

(we call it the induced Markov chain for $\wedge^{(1)}$) and one-step transition probabilities which are the following projections:

$$q_{\alpha\beta}^{\wedge^{(1)}} = \sum_{\beta'} p_{\alpha\beta'} \ , \quad \alpha, \beta \in H_{\wedge^{(1)}} \ ,$$

where the summation is over all β' such that β' belongs to the same face as β and (if this face is $\wedge = \wedge^{(2)}$) the straight line connecting β and β' is perpendicular to $C^{\wedge}_{\wedge^{(1)}}$. From the homogeneity conditions, it follows that the induced chain for $\wedge^{(1)}$ does not depend on the choice of $\alpha \in \wedge^{(1)}$.

Assumption 0_3 For any $\wedge^{(1)}$, the induced chain $\mathcal{L}_{\wedge^{(1)}}$ is irreducible and aperiodic.

Then $\mathcal{L}_{\wedge^{(1)}}$ is ergodic iff $\wedge^{(1)}$ is ergodic. This explains the word. Let $\pi_{\wedge^{(1)}}(h)$ be the stationary probabilities of $\mathcal{L}_{\wedge^{(1)}}$, $h \in H_{\wedge^{(1)}}$, in the ergodic case.
We define, for any ergodic $\wedge^{(1)}$, a number $v_{\wedge^{(1)}}$, the *second vector field* on the one-dimensional ergodic faces

$$v_{\wedge^{(1)}} = \sum_{h \in H_{\wedge^{(1)}}} \pi_{\wedge^{(1)}}(h) Pr_{\wedge^{(1)}} M(h) \ ,$$

where $Pr_{\wedge^{(1)}}$ means *orthogonal projection* of $M(h)$ onto $\wedge^{(1)}$. If $h \in \wedge^{(1)}$, this implies

$$Pr_{\wedge^{(1)}} M(h) = M_{\wedge^{(1)},\wedge^{(1)}} + \sum_{\wedge^{(2)}} Pr_{\wedge^{(1)}} M_{\wedge^{(1)},\wedge^{(2)}} \ .$$

Theorem 5.1.5 *If $v_{\wedge^{(1)}} > 0$ for at least one ergodic $\wedge^{(1)}$, then $\mathcal{L}_T$ is transient.*

Assumption 0_4 $v_{\wedge^{(1)}} \neq 0$ for all $\wedge^{(1)}$.

The sign of $v_{\wedge^{(1)}}$, is easy to calculate, as shown in the following

Lemma 5.1.6

$$\text{sgn}\ (v_{\wedge^{(1)}}) = \text{sgn}\ \Bigg(M_{\wedge^{(1)}}, \wedge^{(1)} + \sum_{\wedge^{(2)}:\wedge^{(1)} \subset \overline{\wedge}^{(2)}} Pr_{\wedge^{(1)}} \left[M_{\wedge^{(1)},\wedge^{(2)}} + M_{\wedge^{(2)}} \frac{Q_{\wedge^{(1)}}^{\wedge^{(2)}} M_{\wedge^{(1)},\wedge^{(2)}}}{Q_{\wedge^{(1)}}^{\wedge^{(2)}} M_{\wedge^{(2)}}} \right] \Bigg) ,$$

where $Q^{\Lambda^{(2)}}_{\Lambda^{(1)}}$ *denotes the projection of a vector in* $\Lambda^{(2)}$ *onto the axis of* $\Lambda^{(2)}$ *other than* $\Lambda^{(1)}$ *and*

$$\operatorname{sgn}(x) = 1 \ \textit{if}\ x > 0 \ \ \textit{and}\ \ -1 \ \textit{if}\ x < 0.$$

5.2 Random walks on hedgehogs

For a given hedgehog $H_{\Lambda^{(1)}}$, we call $C^{\Lambda^{(2)}}_{\Lambda^{(1)}}$ its *bristle*. A bristle $C^{\Lambda^{(2)}}_{\Lambda^{(1)}}$ is called *ingoing* if $\Lambda^{(2)}$ is ingoing, i.e. if the number (representing the mean jump along the bristle)

$$m_{\Lambda^{(2)}} = \sum_{h'} (h' - h)\, q^{\Lambda^{(1)}}_{hh'} \ , \tag{5.1}$$

which does not depend on the position of $h \in C^{\Lambda^{(2)}}_{\Lambda^{(1)}}$, is negative.

When $\Lambda^{(1)}$ is not ergodic, we shall define the *scattering* probability $p_{sc}(\Lambda^{(1)}, \Lambda^{(2)})$, for $\Lambda^{(2)} \in S_-(\Lambda^{(1)})$, which is the probability that the random walk will drift to infinity along $C^{\Lambda^{(2)}}_{\Lambda^{(1)}}$. Under our simplest homogeneity assumptions, this definition does not depend on the initial position, provided that this latter is either at the origin of the hedgehog or on some ingoing bristle. Thus we can assume that it is at the origin $0 \in H_{\Lambda^{(1)}}$.

Computation of the scattering probabilities

Let us fix $\Lambda^{(1)}, \Lambda^{(2)}$ and put $q_{hh'} = q^{\Lambda^{(1)}}_{hh'}$. It is clear that

$$p_{sc}(\Lambda^{(1)}, \Lambda^{(2)}) = \frac{\sum_{h \in C^{\Lambda^{(2)}}_{\Lambda^{(1)}}} q_{0h}\, p(h)}{\sum_{0 \neq h \in H_{\Lambda^{(1)}}} q_{0h}\, p(h)} \ , \tag{5.2}$$

where $p(h) \equiv p_{\Lambda^{(2)}}(h)$ is the probability that, starting from h, the particle on the hedgehog will never return to 0. Formula 5.2 follows from the fact that

$$p_{sc}(\Lambda^{(1)}, \Lambda^{(2)}) = \text{Const} \sum_{h \in C^{\Lambda^{(2)}}_{\Lambda^{(1)}}} q_{0h}\, p(h) \ ,$$

where Const does not depend on the outgoing $\Lambda^{(2)}$. We shall show now that

$$p(h) = 1 - (1 - \gamma)^h \ , \tag{5.3}$$

where $\gamma = p(1)$ is the unique root inside the unit disc of the equation

$$(1 - \gamma)^h = \sum_{h'} q_{hh'}\, (1 - \gamma)^{h'} \ , \tag{5.4}$$

with $h \in C_{\Lambda^{(1)}}^{\Lambda^{(2)}}/\{0\}$ (e.g. we can take $h = 1$), and $h' \in C_{\Lambda^{(1)}}^{\Lambda^{(2)}}$. The proof is easily obtained from the recursive relationship

$$p(h+1) = p(1) + [1 - p(1)]p(h)$$

and standard generating function methods.

Proof of lemma 5.1.6 : Let $H_{\Lambda^{(1)}}$ be a hedgehog such that all $m_{\Lambda^{(2)}}$ are negative, for all its bristles $C_{\Lambda^{(1)}}^{\Lambda^{(2)}}$ .Let

$$\pi_{\Lambda^{(2)}} = \sum_{h \in C_{\Lambda^{(1)}}^{\Lambda^{(2)}} \cap \Lambda^{(2)}} \pi_{\Lambda^{(1)}}(h) .$$

We claim that

$$\frac{\pi_{\Lambda^{(2)}}}{\pi_0} = \frac{Q_{\Lambda^{(1)}}^{\Lambda^{(2)}} \, M_{\Lambda^{(1)},\Lambda^{(2)}}}{Q_{\Lambda^{(1)}}^{\Lambda^{(2)}} \, M_{\Lambda^{(2)}}} .$$

To prove this, we note first that, for computing the above quantity, it suffices to consider a modified random walk on the bristle $C_{\Lambda^{(1)}}^{\Lambda^{(2)}}$, i.e. on $\mathbf{Z}_+^1$, after slightly *updating* the transition probabilities. More exactly, we define $\tilde{q}_{hh'} = q_{hh'}$, for all $h, h' \in C_{\Lambda^{(1)}}^{\Lambda^{(2)}}$, except for $\tilde{q}_{00}$ which is taken equal to

$$\tilde{q}_{00} = 1 - \sum_{0 \neq h' \in C_{\Lambda^{(1)}}^{\Lambda^{(2)}}} \tilde{q}_{0h'} .$$

Then $\pi_{\Lambda^{(2)}}/\pi_0$ does not depend on this modification and its value in the case of $\mathbf{Z}_+^1$ is a well known result, yielding in particular exact ergodicity conditions for random walks in $\mathbf{Z}_+^2$. (The point is that, due to the homogeneity, it is not necessary to compute the exact values of the $\pi_{\Lambda^{(1)}}(h)$s .Only the drifts are needed.)
The proof is concluded. ■

Remark We have solved an *exit boundary* problem, using terminology from Martin's theory. See Feller [Fel56] for instance.

5.3 Formulation of the main result

Definition 5.3.1 *For given T and $\mathcal{L}_T$, we define the following associated Markov chain $\mathcal{M}$ having a finite number of states $n = | T |$, equal to the number of two-dimensional faces of T. (It is thus natural to denote these*

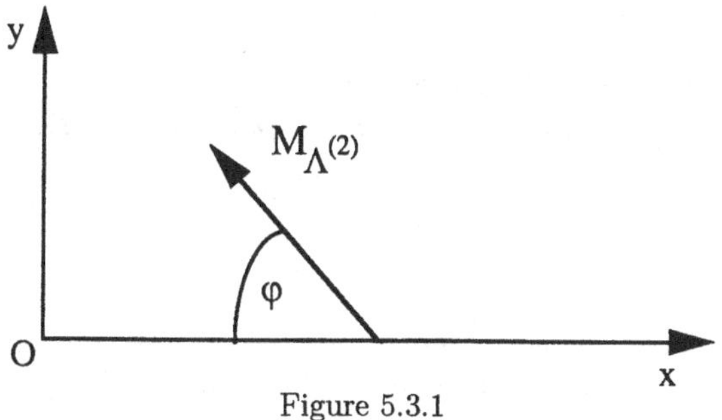

Figure 5.3.1

states by $\wedge^{(2)}$.) *The one-step transition probabilities* $p(\wedge_i^{(2)}, \wedge_j^{(2)})$ *of* $\mathcal{M}$ *are equal to*

$$p(\wedge_i^{(2)}, \wedge_j^{(2)}) = \begin{cases} p_{sc}(\wedge^{(1)}, \wedge_j^{(2)}), \ \text{if } \wedge_i^{(2)} \in S_+(\wedge^{(1)}) \ , \ \wedge_j^{(2)} \in S_-(\wedge^{(1)}), \\ 1, \ \text{if } \wedge_i^{(2)} = \wedge_j^{(2)} \in S_+(\wedge^{(1)}) \text{ for some ergodic } \wedge^{(1)}, \\ 0, \ \text{otherwise}. \end{cases}$$

We do not exclude that the associated chain be reducible or periodic. Let $\mathfrak{A}_1, \ldots, \mathfrak{A}_k$ be irreducible classes of essential states. Let us consider some class $\mathfrak{A}_i$ with $|\mathfrak{A}_i| \geq 2$. We define the following function f on $\mathfrak{A}_i$: if $\wedge^{(2)} \in \mathfrak{A}_i$ and $\phi_{\wedge^{(2)}}$ is the angle between $M_{\wedge^{(2)}}$ and the negative axis from which $M_{\wedge^{(2)}}$ goes away, as shown in fig. 5.3.1, then we put

$$f(\wedge^{(2)}) = \log \tan(\phi_{\wedge^{(2)}}) \, .$$

Heuristically, if we are e.g. on the x-axis of $\wedge^{(2)}$ at a point $(x, 0)$ and we move along the constant vector field $M_{\wedge^{(2)}}$ to a point $(0, y)$ of the y-axis, then $\exp f(\wedge^{(2)})$ represents the dilatation coefficient $\alpha = y/x$.

- If $\mathfrak{A}_i$ is aperiodic and $\pi_i(\wedge^{(2)})$ denotes its stationary probability, we define

$$L(\mathfrak{A}_i) = \sum_{\wedge^{(2)} \in \mathfrak{A}_i} \pi_i(\wedge^{(2)}) f(\wedge^{(2)}) \, . \tag{5.5}$$

- For a periodic $\mathfrak{A}_i$, $\pi_i(\wedge^{(2)})$ is then taken to be the stationary probability in the aperiodic subclass containing $\wedge^{(2)}$.

The vector $M(\mathfrak{A}_i)$ is defined by the same formula (5.5). Perhaps it will be more convenient to normalize it, multiplying by $N(\mathfrak{A}_i)^{-1}$ where $N(\mathfrak{A}_i)$ is the number of aperiodic subclasses in $\mathfrak{A}_i$.

Assumption 0$_5$ For all i , $L(\mathfrak{A}_i) \neq 0$.

Our main result is the following:

Theorem 5.3.2 *Under the assumptions* $0_1, \ldots, 0_5$ *and if the assumptions of theorems 5.1.2 and 5.1.5 are not fulfilled, then* $\mathcal{L}_T$ *is recurrent if, and only if, for any* $\mathfrak{A}_i$ *with* $|\mathfrak{A}_i| \geq 2$,

$$L(\mathfrak{A}_i) < 0 \,. \tag{5.6}$$

It appears that (5.6) is not sufficient for the chain to be ergodic and in fact we will find both ergodicity and null recurrence regions. To that end, we have to define a new important quantity $M(\mathfrak{A}_i)$, connected with $\mathfrak{A}_i$. Let us denote by $\wedge_0^{(2)}, \ldots, \wedge_t^{(2)}, \ldots$ the random states of $\mathfrak{A}_i$ such that $\wedge_0^{(2)} \in \mathfrak{A}_i$.

Lemma 5.3.3 *The limit*

$$\lim_{t\to\infty} \frac{1}{t} \log(E \prod_{i=1}^{t} \tan \phi_{\wedge_i}) \stackrel{\text{def}}{\equiv} M(\mathfrak{A}_i) \tag{5.7}$$

exists and does not depend on $\wedge_0^{(2)} \in \mathfrak{A}_i$. *Moreover*

$$M(\mathfrak{A}_i) = \log \lambda_1(\mathfrak{A}_i) \tag{5.8}$$

where λ_1 *is the maximal eigenvalue of the* $n_i \times n_i$*- matrix,* $n_i = |\mathfrak{A}_i|$, *with matrix elements*

$$A(\wedge_i^{(2)}, \wedge_j^{(2)}) = p(\wedge_i^{(2)}, \wedge_i^{(2)}) \sqrt{\tan \phi_{\wedge_i^{(2)}} \tan \phi_{\wedge_j^{(2)}}} \,, \ \textit{for}\ \wedge_i^{(2)},\ \wedge_j^{(2)} \in \mathfrak{A}_i. \tag{5.9}$$

Proof of Lemma 5.3.3 : Let us note first that for any two vectors l_1, l_2 with positive components

$$\lim_{N\to\infty} \frac{1}{N} \log(l_1, A^N l_2) = \log \lambda_1$$

and then we notice that (5.7) can be represented in such a way.

Assumption 0_6 For all i, $|\mathfrak{A}_i| \geq 2$,

$$M(\mathfrak{A}_i) \neq 0 \,. \tag{5.10}$$

In the following theorem we assume also that all states of $\mathcal{M}$ are essential. In general, the limit (5.7) can depend on the initial inessential state. All the proofs however are completely the same.

Theorem 5.3.4 *Under the assumptions* $0_1, \ldots, 0_6$ *and if the assumptions of theorems 5.1.2 and 5.1.5 are not fulfilled, then* $\mathcal{L}_T$ *is ergodic if, and only if, for any* $\mathfrak{A}_i$ *with* $|\mathfrak{A}_i| \geq 2$,

$$L(\mathfrak{A}_i) < 0 \text{ and } M(\mathfrak{A}_i) < 0. \tag{5.11}$$

This implies that, if for all i we have (5.6), but for at least one i we have

$$M(\mathfrak{A}_i) > 0,$$

then $\mathcal{L}_T$ *is null recurrent.*

Remark Because of the inequality

$$\log E\xi \geq E \log \xi,$$

valid for any positive r.v. ξ, we always have

$$L(\mathfrak{A}_i) \leq M(\mathfrak{A}_i). \tag{5.12}$$

In particular, if

$$M(\mathfrak{A}_i) < 0,$$

then

$$L(\mathfrak{A}_i) < 0.$$

The practical computation of the ergodicity conditions can be achieved according to the following sequence of steps:

(i) calculate the vectors of mean jumps;

(ii) calculate sgn $v_{\wedge^{(1)}}$ for all ergodic $\wedge^{(1)}$, using lemma 5.1.6;

(iii) calculate the scattering probabilities, using formulas (5.2)–(5.4);

(iv) calculate the stationary probabilities of the associated chain, which in the general case give rise to a system of $|T|$ linear equations;

(v) calculate $M(\mathfrak{A}_i)$ using (5.8);

(vi) use theorems 5.1.2, 5.1.5, 5.3.4.

So we get a complete classification up to the assumptions $0_1, \ldots, 0_6$.

5.4 Quasi-deterministic process

Here, we introduce an auxiliary process η_t on $\tilde{T}$ with $t \in R_+$. Consider a particle η_t moving along the constant vector field $M_{\Lambda^{(2)}}$ (with velocity $M_{\Lambda^{(2)}}$) on any face $\tilde{\Lambda}^{(2)} \subset \tilde{T}$. When it reaches a one-dimensional face $\tilde{\Lambda}^{(1)}$, it chooses with probability $p(\Lambda^{(1)}, \Lambda_i^{(2)})$ a face $\Lambda_i^{(2)} \in S_-(\Lambda^{(1)})$ and continues its way along $\tilde{\Lambda}_i^{(2)}$ and so on. Thus η_t is deterministic outside one-dimensional faces. Let $\eta_0 = x \in \tilde{T}$ and let

$$\tau_1(x) < \tau_2(x) < \dots$$

be all the instants when η_t is on a one-dimensional face of $\tilde{T}$. Let us define the discrete time embedded process

$$\chi_n = \chi_n(x) \equiv \eta_{\tau_n(x)} \cdot$$

We shall find the conditions of ergodicity, null recurrence and transience for the process η_t, which will appear to be the same as for the corresponding random walk.

Lemma 5.4.1 *Under the assumptions of theorem 5.3.2, the following conditions are equivalent:*

(i) *(5.6) holds;*
(ii) *for any x, $\chi_n(x)$ reaches a.s. any neighbourhood of 0;*
(iii) *for any x, the time for $\eta_t(x)$ to reach any neighbourhood of 0 is a.s. finite.*

Proof If (5.6) holds, then $\log \chi_n \to -\infty$ a.s., by the strong law of large numbers. So (i)$\Longrightarrow$ (ii) $\Longleftrightarrow$ *(iii)*. Conversely, if we have

$$L(\mathfrak{A}_i) > 0,$$

then $\log \chi_n \to \infty$ a.s., so that, with probability 1, η_t cannot reach any neighbourhood of 0 . ■

Lemma 5.4.2 *Under the assumptions of theorem 5.3.2 the following conditions are equivalent:*

(i) *(5.11) holds;*
(ii) *the $E\tau_n$s are uniformly bounded, i.e. $\sum_{i=1}^{\infty} E(\tau_i - \tau_{i-1}) < \infty$, $\tau_0 = 0$;*
(iii) *$\eta_t(x)$ reaches zero in a finite mean time;*
(iv) *$\eta_t(x)$ reaches any neighbourhood of 0 in a finite meantime.*

Proof Obviously we have

$$\text{(ii)} \Longleftrightarrow \text{(iii)} \Longrightarrow \text{(iv)}$$

Let us prove (i) $\Longrightarrow$ (ii). In fact, if (5.9) holds, then $E(\chi_n)$ exponentially converges to zero, so that

$$E(\chi_n) \leq Ce^{-\alpha n},$$

for some $C > 0$, $\alpha > 0$ and all n. But also

$$E(|\tau_n - \tau_{n-1})| \leq CE(\chi_n) .$$

So we have proved that (i) implies (ii). Let us now prove that (iv) $\Longrightarrow$ (i), i.e. if

$$L(\mathfrak{A}_i) > 0,$$

then the mean time for $\eta_t(x)$ to reach a neighbourhood of 0 is infinite. Define now

$$\tilde{T}(\mathfrak{A}_i) = \bigcup_{\wedge \in \mathfrak{A}_i} \overline{\wedge} .$$

For any $x \in \tilde{T}(\mathfrak{A}_i)$ and any $\epsilon > 0$, let $\tau^\epsilon(x)$ be the time of first reaching the set $\{y \in \tilde{T}(\mathfrak{A}_i) : \| y \| \leq \epsilon\}$ by the quasi-deterministic process $\eta_t(x)$ (we take $\tau^\epsilon(x) = 0$ if $\| X \| \leq \epsilon$).

Proposition 5.4.3 *Let $x \in \tilde{T}(\mathfrak{A}_i)$ and for any $\epsilon > 0$*

$$E(\tau^\epsilon(x)) < \infty.$$

Then for any $r > 0$ and for any $\epsilon > 0$, we have

$$\frac{1}{r}E(\tau^{r\epsilon}(x)) = E(\tau^\epsilon(x)) . \tag{5.13}$$

This is the obvious scaling property of the process η_t. ■

Proposition 5.4.4 *If there exists $x \in \tilde{T}(\mathfrak{A}_i)$ such that for any $\epsilon > 0$*

$$E(\tau^\epsilon(x)) < \infty ,$$

then this property holds for any x.

Proof Let us fix such x. Then, for any $y \in \tilde{T}(\mathfrak{A}_i)$, there exist $t \in R_+$ and $r \in R_+$ such that

$$\eta_t(x) = ry ,$$

with a positive probability. It follows that, for any $\epsilon > 0$,

$$E(\tau^{\epsilon}(ry)) < \infty,$$

so that by proposition 5.4.3,

$$E(\tau^{\epsilon}(y)) < \infty,$$

for any $\epsilon > 0$. ■

Proposition 5.4.5 *If there exists $x \in \tilde{T}(\mathfrak{A}_i)$ such that, for any $\epsilon > 0$,*

$$E(\tau^{\epsilon}(x)) < \infty \,, \tag{5.14}$$

then, for any $x \in \tilde{T}(\mathfrak{A}_i)$,

$$E(\tau^{0}(x)) < \infty,$$

where $\tau^0(x)$ is the first time of reaching 0 for the process $\eta_t(x)$.

Proof By proposition 5.4.4, equation (5.14) holds for all x and $\epsilon > 0$. As the number of one-dimensional faces is finite, it follows from proposition 5.4.3 that, for any $\epsilon > \epsilon' > 0$, we have

$$\sup_{x \in \tilde{T}(\mathfrak{A}_i)\,:\,\|x\|=\epsilon} E(\tau^{\epsilon'}(x)) < \infty.$$

Let us consider the sequence $\epsilon_n = 1/2^n$, $n = 0, 1, 2, \ldots$. Let

$$\sup_{x \in \tilde{T}(\mathfrak{A}_i)\,:\,\|x\|=\epsilon_0} E(\tau^{\epsilon_1}(x)) = C_1 < \infty.$$

Then, for any n, by proposition 5.4.3,

$$\sup_{x\,:\,\|x\|=\epsilon_n} E(\tau^{\epsilon_{n+1}}(x)) = C_1 2^{-n}. \tag{5.15}$$

Let now $x \in \tilde{T}(\mathfrak{A}_i)$, $\| x \| = 1$, and put

$$t^1(x) = \tau^{\epsilon_1}(x), \ldots, \; t^{k+1}(x) = t^k(x) + \tau^{\epsilon_{k+1}}(\eta_{t^k(x)}(x)), \ldots.$$

First, we have

$$t^k(x) \uparrow \tau^0(x) \text{ a.s., for } \; k \to \infty, \tag{5.16}$$

which in fact corresponds to the definition of $\tau^0(x)$. Secondly, by (5.15),

$$E(t^k(x)) \leq \sum_{j=1}^{k} \sup_{y \in \tilde{T}(\mathfrak{A}_i)\,:\,\|y\|=\epsilon_j} E(\tau^{\epsilon_{j+1}}(y)) \leq C_1 \sum_{j=1}^{k} (\frac{1}{2})^k.$$

Hence,

$$E(\tau^0(x)) < \infty,$$

for all x, as in proposition 5.4.3.
Proposition 5.4.5 is proved. ■

But at the same time, we see, as above, that, if $M(\mathfrak{A}_i) > 0$, then $E(\chi_n(x))$ increases exponentially fast, so that $\tau^0(x) = \infty$. We reach a contradiction which proves the assertion (iv) $\Longrightarrow$ (i) and also lemma 5.4.2. ■

5.5 Proof of the ergodicity in theorem 5.3.4

Let us first consider the case when there is a single essential class with at least two essential states. So, no one-dimensional face is ergodic. The process η_t is called *ergodic* if the mean time f_x of reaching 0, starting from a point $x \in \tilde{T}$, is finite (for all x). We would like to use f_x as a Lyapounov function for $\mathcal{L}_T$ in the following criterion for ergodicity (see chapter 2): $\mathcal{L}_T$ is ergodic if and only if there exist a positive integer-valued function $m(\alpha)$ and $\epsilon > 0$, such that

$$\sum_\beta \tilde{p}_{\alpha\beta}^{(m(\alpha))} f_\beta - f_\alpha < -\epsilon\, m(\alpha)\,, \tag{5.17}$$

for $\alpha \in T - T_0$ and some finite set $T_0 \subset T$. Let $\tilde{p}_{\alpha\beta}^{(t)}$ be the transition probabilities in t steps of the process η_t for $\alpha, \beta \in \tilde{T}$. For α and t given, they differ from zero only for a finite number of points β.
Suppose $M(\mathfrak{A}_i) < 0$. Then η_t is ergodic and therefore, for all α not very close to the origin,

$$f_\alpha = 1 + \sum_\beta \tilde{p}_{\alpha\beta}^{(1)} f_\beta$$

or

$$\sum_\beta \tilde{p}_{\alpha\beta}^{(1)} f_\beta - f_\alpha = -1\,. \tag{5.18}$$

We note that f_α has the following properties:

(i) it is continuous everywhere, except on the one-dimensional faces where scattering occurs;
(ii) $C_1 \mid x \mid \leq f_x \leq C_2 \mid x \mid$, for some C_1, $C_2 > 0$;
(iii) for any two-dimensional face $\Lambda^{(2)}$, the function f_x has a linear decrease along any line parallel to $M_{\Lambda^{(2)}}$ in the direction of $M_{\Lambda^{(2)}}$;

(iv) due to the space homogeneity, f_x satisfies (5.13) with $m(\alpha) = 1$ everywhere, except possibly in some neighbourhoods of one-dimensional faces. More precisely, let us choose some non-ergodic one-dimensional face $\tilde{\Lambda}^{(1)}$ and put, for any $\tilde{\Lambda}^{(2)}$,

$$\mathcal{D}_{\Lambda^{(1)},\Lambda^{(2)}}(\rho) = \{\alpha : \alpha \in \overline{\Lambda^{(2)}}, \rho(\alpha, \tilde{\Lambda}^{(1)}) = \rho\}.$$

Then, if $\Lambda^{(2)} \in S_-(\Lambda^{(1)}), \alpha \in \mathcal{D}_{\Lambda^{(1)},\Lambda^{(2)}}(1)$, (5.17) might not be satisfied. For this reason, we modify our Lyapounov functions as follows. Define

$$\tilde{f}_\alpha = \begin{cases} f_\alpha \, , \; \alpha \notin \mathcal{D}(\rho_0) \, , \\ (C_2 + 1) \mid \alpha \mid , \; \alpha \in \mathcal{D}(\rho_0) \, , \end{cases}$$

where

$$\mathcal{D}(\rho_0) = \bigcup_{\Lambda^{(1)}} \bigcup_{\Lambda^{(2)} \in S_-(\Lambda^{(1)})} \bigcup_{\rho=0}^{\rho_0} \mathcal{D}_{\Lambda^{(1)},\Lambda^{(2)}}(\rho)$$

and ρ_0 is a constant to be specified below.

Lemma 5.5.1 *There exist $\rho_0, m, \delta > 0$, such that (5.17) holds for the new Lyapounov function $\tilde{f}_\alpha$, with*

$$m(\alpha) = \begin{cases} m, & \alpha \in \mathcal{D}(\rho_0 + 1) \, , \\ \delta \mid \alpha \mid, & \alpha \in \bigcup_{\Lambda^{(1)}} \bigcup_{\Lambda^{(2)} \in S_+(\Lambda^{(1)})} \mathcal{D}_{\Lambda^{(1)},\Lambda^{(2)}}(1) \, , \\ 1, & \text{in other cases.} \end{cases}$$

Proof We choose $\delta > 0$ sufficiently small ; then we take ρ_0 sufficiently large and then $m = m(\rho_0)$ sufficiently large. When $m(\alpha) = 1$, it is easy to verify (5.17), due to the linearity property (iv).

Let us fix now a point $\alpha \in \mathcal{D}(\rho_0 + 1)$. One can prove that starting from α after m steps, we shall be outside $\mathcal{D}(\rho_0)$ with probability $1 - \epsilon_1$, where $\epsilon_1 = \epsilon_1(m) \to 0$ when $m \to \infty$, uniformly in α, with $\mid \alpha \mid > a_0$, for some a_0 sufficiently large. This follows just from the transience of the corresponding hedgehog. It follows also that (5.17) holds since, for large m, we can take ϵ_1 arbitrarily small with

$$\tilde{f}(\beta) = f(\beta) < C_2(\mid \alpha \mid + md) \, ,$$

where $\beta \notin \mathcal{D}(\rho_0)$ is the final point after m steps.

Let now $\alpha \in \mathcal{D}_{\Lambda^{(1)},\Lambda^{(2)}}(1)$, $\Lambda^{(2)} \in S_+(\Lambda^{(1)})$ and $\xi_t(\alpha)$ be the position of the random walk starting from α.

Let $\wedge_i \in S_-(\wedge')$. Denote by $\eta_t(\alpha, \wedge_i)$ the point of $\mathbf{R}_+^2 = \wedge_i$ which is the unique point of $\wedge_i$, where the process η_t will be found at time t, after having started from α. We need the following intermediate result.

Lemma 5.5.2 *Let $t = \delta \mid \alpha \mid$. Then, for any $\epsilon_2, \epsilon_3 > 0$ sufficiently small, there exists $a_0 > 0$ such that, for any $\mid \alpha \mid > a_0$ and for any $\wedge_i \in S_-(\wedge^{(1)})$,*

$$\mid \mathrm{P}\left(\xi_t(\alpha) \in \wedge_i, \mid \xi_t(\alpha) - \eta_t(\alpha, \wedge_i) \mid < \epsilon_3 \mid \alpha \mid\right) - p(\wedge, \wedge_i) \mid < \frac{\epsilon_2}{l}, \quad (5.19)$$

where $l = |S_-(\wedge^{(1)})|$.

Proof From the point $\alpha \in \mathcal{D}_{\wedge',\wedge}(1)$, we first make $\frac{\epsilon_3}{2(d+1)}|\alpha|$ jumps. Then, for $|\alpha|$ large enough, with probability p_i such that

$$\mid p_i - p(\wedge, \wedge_i) \mid < \frac{\epsilon_2}{3l},$$

we shall reach a point $\alpha_i \in \wedge_i$ satisfying

$$\rho(\alpha_i, \eta_t(\alpha, \wedge_i)) < \frac{1}{2}\epsilon_3|\alpha|.$$

After having started from α_i, we perform the remaining

$$(\delta - \frac{\epsilon_3}{2(d+1)}) \mid \alpha \mid$$

jumps. This will be in fact a translation invariant random walk in $\mathbf{Z}^2$ and, using Kolmogorov's inequality, we prove that, for $|\alpha|$ large, it will never go out of $\wedge_i$ with probability $1 - \frac{\epsilon_2}{3l}$. Moreover, by the law of large numbers, the final point $\xi_t(\alpha_i, \alpha)$ of this random walk satisfies the inequality

$$\xi_t(\alpha_i, \alpha) - (\alpha_i + M_{\wedge_i}(\delta - \frac{\epsilon_3}{2(d+1)}) \mid \alpha \mid) < \frac{1}{2}\,\epsilon_3 \mid \alpha \mid,$$

with probability $1 - (\epsilon_2/3l)$. Putting together all these estimates, we get (5.19), concluding by the way the proof of lemma 5.5.2 ■

We are now in a position to finish the proof of lemma 5.5.1.
From (5.18), we get

$$\sum_\beta \tilde{p}_{\alpha\beta}^{(t)}\, \tilde{f}_\beta - \tilde{f}_\alpha = -t\,. \quad (5.20)$$

Comparing (5.17) and (5.20) yields

$$\sum p_{\alpha\beta}^{(t)}\, \tilde{f}_\beta - \tilde{f}_\alpha = \sum_\beta \tilde{p}_{\alpha\beta}^{(t)}\, \tilde{f}_\beta - \tilde{f}_\alpha + \Delta\,,$$

where

$$\Delta < \epsilon_2\,(C_2+1)|\alpha|(1+d\delta)+\epsilon_3|\alpha|\,. \tag{5.21}$$

Thus, for ϵ_2, ϵ_3 sufficiently small, we get

$$\sum p_{\alpha\beta}^{(t)}\,\tilde{f}_\beta - \tilde{f}_\alpha < -\frac{1}{2}\,\delta|\alpha|\,.$$

This concludes the proof of lemma 5.5.1 and of the ergodicity. ■

If there are several essential classes, we use the same Lyapounov function as before inside two-dimensional faces and in a vicinity of non-ergodic one-dimensional faces. We define it in a neighbourhood of ergodic faces, exactly as it was done in chapter 4 for $\mathbf{Z}_+^3$.

5.6 Proof of the transience

Assume that, for the class $\mathfrak{A}_i$,

$$L(\mathfrak{A}_i) > 0\,. \tag{5.22}$$

We shall then prove that $\mathcal{L}_T$ is transient.

Let $\tilde{\xi}_t(\alpha)$ be the position of the random walk corresponding to $\mathcal{L}$ with the initial condition α, i.e. $\tilde{\xi}_0(\alpha)=\alpha$. Choosing $\alpha \neq \underline{0}$ belonging to some $\Lambda^{(1)}$, we define by induction the sequence of random times $0=\tau_0<\tau_1<\ldots<\tau_n<\ldots$ such that

(a) if $\xi_{\tau_{n-1}}(\alpha)\neq 0, \tau_n$ is the first hitting time (after τ_{n-1}) of $\underline{0}$ or of some one-dimensional face different from the face $\Lambda_{n-1}^{(1)}$ to which $\tilde{\xi}_{\tau_{n-1}}(\alpha)$ belongs;

(b) if $\tilde{\xi}_{\tau_n}(\alpha)=0$ for some τ_n, then $\tau_{n+1}=\tau_n+1$.

Let us consider the new Markov chain $\zeta_n(\alpha) = \tilde{\xi}_{\tau_n}(\alpha)$, $\zeta_0(\alpha) = \alpha$, the state space of which is the union of all one-dimensional faces $\bigcup\Lambda^{(1)}$ and $\underline{0}$. The probability of sometimes hitting $\underline{0}$ is identical for $\xi_n(\alpha)$ and $\zeta_n(\alpha)$. So it is sufficient to prove the non-recurrence of $\zeta_n(\alpha)$.

Lemma 5.6.1 *Let us consider two one-dimensional faces $\Lambda^{(1)}, \Lambda_1^{(1)}$, and the corresponding two-dimensional ones*

$$\Lambda_1^{(2)} \in S_+(\Lambda^{(1)}),\ \Lambda^{(2)} \in S_-(\Lambda^{(1)}) \cap S_+(\Lambda_1^{(1)})\,.$$

Then, for any $\epsilon > 0$, *one can find* $D_1 > 0$ *such that, for any* $\beta \in \Lambda^{(1)}$,

$$\left| P\left(\zeta_1(\beta) \in \Lambda_1^{(1)}, \left| |\zeta_1(\beta)| - \tan \, \phi_{\Lambda_1^{(2)}} \right| \le \epsilon|\beta|\right) - p(\Lambda_1^{(2)}, \Lambda^{(2)}) \right| \le \frac{D_1}{|\beta|} . \tag{5.23}$$

Proof Let us note first that, for any $\Lambda_2^{(2)} \in S_+(\Lambda^{(1)})$, we have

$$p(\Lambda_1^{(2)}, \Lambda^{(2)}) = p(\Lambda_2^{(2)}, \Lambda^{(2)}) .$$

As the jumps are bounded, we have $\tau_1 \ge [|\beta|/d]$ a.s. It follows (see theorem 2.1.7) that there exist $\delta_1, \delta_2, D_2 > 0$, such that, for any $\beta \in \Lambda^{(1)}$ and any $t \in \mathbf{Z}_+, t < [|\beta|/d]$,

$$P\left(\rho(\check{\xi}_t(\beta), \Lambda^{(1)}) < \delta_1 t\right) \le D_2 \, e^{-\delta_2 t} .$$

We also have

$$\left| P\left(\rho(\check{\xi}_t(\beta), \Lambda^{(1)}) > \delta_1 t, \, \check{\xi}_t(\beta) \in \Lambda^{(2)}\right) - p(\Lambda_1^{(2)}, \Lambda^{(2)}) \right| \le C_3 \, e^{-\delta_2 t} , \tag{5.24}$$

for any $t < |\beta|/d$, where C_3 is a positive constant.
For any ϵ_1, where $0 < \epsilon_1 < 1$, and $t_1 = [\epsilon_1 |\beta|/d]$, we have

$$|\check{\xi}_{t_1}(\beta) - \beta| \le \epsilon_1 |\beta| . \tag{5.25}$$

To prove (5.23), it is sufficient, by (5.24), (5.25), to prove that, for $\check{\xi}_{t_1}(\beta)$ such that

$$\rho(\check{\xi}_{t_1}(\beta), \Lambda^{(1)}) > \delta_1 \, t_1 ,$$

and for any $\epsilon_2 > 0$, there exists a constant $C_4 = C_4(\epsilon_1, \epsilon_2) > 0$ (not depending on β) such that

$$\begin{aligned} P(\check{\xi}_{\tau_1}(\beta) \in \Lambda_1^{(1)}, \rho(\check{\xi}_{\tau_1}(\beta), \check{\xi}_{t_1}(\beta) + (\tau_1 - t_1) M_{\Lambda^{(2)}}) \\ > \epsilon_2 |\beta| \, \Big| \, \rho(\check{\xi}_{t_1}(\beta), \Lambda^{(1)}) > \delta_1 t_1) \le \frac{C_4}{\beta} . \end{aligned} \tag{5.26}$$

To prove (5.26), we consider $\Lambda^{(2)}$ embedded into $\mathbf{Z}^2$ and the space homogeneous r.w. ξ_t', $t \in \mathbf{Z}_+$, on this $\mathbf{Z}^2$, with the initial position

$$\xi_0' = \check{\xi}_{t_1}(\beta)$$

and one-step transition probabilities

$$p'(\alpha, \beta) = p'(\alpha + \alpha', \beta + \alpha') = p(\alpha, \beta) ,$$

for all $\alpha, \beta \in \Lambda^{(2)}$, $\alpha' \in \mathbf{Z}^2$.

Let τ' be the first hitting time of $\wedge_1^{(1)}$ by ξ'_t. We can choose $T \in \mathbf{Z}_+$ and $\epsilon_3 > 0$, so that

$$\rho(\xi'_0, \wedge^{(1)}) \geq \delta_1 t = \delta_1 \left[\frac{\epsilon_1 |\beta|}{d} \right] > \epsilon_3 |\beta|$$

and

$$\begin{aligned} &\rho\,(\xi'_0, \wedge_1^{(1)}) + TPr_{\wedge^{(1)}} M_{\wedge^{(2)}} + \epsilon_3 |\beta| \\ &\quad \leq |\beta|(1+\epsilon_1) + TPr_{\wedge^{(1)}}\, M_{\wedge^{(2)}} + \epsilon_3|\beta| \ < 0\,. \end{aligned}$$

Let us note that, if

$$\max_{t \in [0,T]} |\xi'_t - \xi'_0 - M_{\wedge^{(2)}}\, t| \leq \epsilon_3 |\beta|\,,$$

then $\tau' < T$ and $\xi'_t \in \wedge^{(2)}$, for all $t < \tau'$. From this remark and Kolmogorov's inequality, we obtain

$$\begin{aligned} &P\left(\breve{\xi}_{\tau_1}(\beta) \in \wedge_1^{(1)}, \rho(\breve{\xi}_{\tau_1}(\beta), \beta_1 + M_{\wedge^{(2)}}(\tau_1 - t_1)) \leq \epsilon_3 |\beta| \,\Big|\, \breve{\xi}_{t_1}(\beta)\right) \\ &\geq P\left(\max_{t \in [t_1, \tau_1]} |\breve{\xi}_t(\beta) - \breve{\xi}_{t_1}(\beta) - (t - t_1) M_{\wedge^{(2)}}| \ \leq \epsilon_3 |\beta| \Big| \breve{\xi}_{t_1}(\beta)\right) \\ &= P\left(\max_{t \in [0,\tau']} |\xi'_t - \xi_0 - t M_{\wedge^{(2)}}| \ \leq \epsilon_3\ |\beta| \right) \\ &\geq P\left(\max_{t \in [0,T]} |\xi'_t - \xi_0 - t M_{\wedge^{(2)}}| \ \leq \epsilon_3 |\beta| \right) \\ &\geq 1 - \frac{C_5 T}{\epsilon_3^2 |\beta|^2}\,, \end{aligned}$$

for some constant $C_5 > 0$. The last step to derive (5.26) is achieved by choosing T such that

$$T \leq Const\ |\beta|.$$

Lemma 5.6.1 is proved. ■

It follows from lemma 5.6.1 that, for any sequence $\wedge_0, \ldots, \wedge_n \in \mathfrak{A}_i$ of two-dimensional faces, for any $\alpha \in \overline{\wedge}_0 \cap \overline{\wedge}_1$ and any $\epsilon > 0, n \in \mathbf{Z}_+$, there exists $C_6 = C_6(\epsilon, n)$ not depending on $\alpha \in \overline{\wedge}_0 \cap \overline{\wedge}_1$, such that, setting

$$\begin{aligned} U &= P\{\zeta_k(\alpha) \in \overline{\wedge}_k \cap \overline{\wedge}_{k+1}, |\zeta_k(\alpha)| \\ &\quad \geq (\tan \phi_{\wedge_k} - \epsilon)|\zeta_{k-1}(\alpha)|, k = 1, \ldots, n\}, \\ V &= p(\wedge_0, \wedge_1) p(\wedge_1, \wedge_2) \ldots p(\wedge_{n-1}, \wedge_n)\,, \end{aligned}$$

we have

$$|U - V| \leq \frac{C_6}{|\alpha|}\,. \tag{5.27}$$

Let us choose $\epsilon > 0$, $\delta > 0$ and $n \in \mathbf{Z}_+$, so that, for any $\Lambda_0 \in \mathfrak{A}_i$,

$$\sum_{k=1}^{n} \sum_{\Lambda^{(2)}} p^{(k)}(\Lambda_0, \Lambda^{(2)}) \log(\tan \phi_{\Lambda^{(2)}} - \epsilon) > \delta > 0 . \tag{5.28}$$

Due to (5.22), it is always possible to satisfy (5.28), since

$$\frac{1}{n} \sum_{k=1}^{n} p^{(k)}(\Lambda_0, \Lambda^{(2)}) \to \pi\ (\Lambda^{(2)})\ , \text{ as } n \to \infty ,$$

for any $\Lambda_0, \Lambda^{(2)}$, whenever Λ_0 is an essential state of the associated Markov chain.

To prove the transience of $\zeta_k(\alpha)$, $k \in \mathbf{Z}_+$, it is sufficient to prove the transience of the chain $\eta_k = \zeta_{nk}(\alpha)$, $k \in \mathbf{Z}_+$. We shall use Foster's criterion (see chapter 4) recalled here: if for a Markov chain with state space X and transition probabilities p_{ij} , $i, j, \in X$, there exist a positive function f defined on the state space X and a set $A \subset X$, such that

$$\begin{cases} \sum_{j \in X} p_{ij}\, f_j \le f_i\ , & \forall i \in X - A\ , \text{ and} \\ \inf_{i \in A} f_i > \sup_{j \in X - A} f_j\ , \end{cases}$$

then the Markov chain is transient. We shall define f on the state space of η_k.

$$f(\alpha) = \begin{cases} \dfrac{1}{\log 3|\alpha|}\ , & \text{if } \alpha \neq 0 \text{ and } \alpha \in \overline{\Lambda}^{(2)}\ , \text{ for some } \Lambda^{(2)} \in \mathfrak{A}_i\ , \\ 1, & \text{in the other cases.} \end{cases}$$

Let us prove that, if $|\eta_0|$ is sufficiently large, then

$$E(f(\eta_1)) \le f(\eta_0)\ . \tag{5.29}$$

In fact, if $f(\eta_0) = 1$, then (5.29) evidently holds. Let now

$$f(\eta_0) = \frac{1}{\log 3|\eta_0|} \text{ and } \eta_0 \in \Lambda^{(1)}\ ,$$

for some one-dimensional face $\Lambda^{(1)}$. Then, by (5.27),

$$E(f(\eta_1)) \le$$

$$\sum_{\Lambda_1, \dots, \Lambda_n} p(\Lambda_0, \Lambda_1) \cdots p(\Lambda_{n-1}, \Lambda_n) \frac{1}{\log[3|\eta_0| \prod_{j=1}^{n} (\tan \phi_{\Lambda_j} - \epsilon)]} + \frac{C_6}{|\eta_0|}, \tag{5.30}$$

where $\Lambda_0 \in S_-(\Lambda^{(1)})$, $\Lambda_0 \in \mathfrak{A}_i$. But, for $|\eta_0|$ sufficiently large,

$$\left[\log \left[3|\eta_0| \prod_{j=1}^{n} (\tan \phi_{\Lambda_j} - \epsilon) \right] \right]^{-1}$$

$$= \frac{1}{\log 3|\eta_0|}\left[1+\frac{1}{\log 3|\eta_0|}\left(\sum_{j=1}^{n}\log(\tan\ \phi_{\wedge_j}-\epsilon)\right)\right]^{-1}$$

$$\leq \frac{1}{\log 3|\eta_0|}-\frac{\sum_{j=1}^{n}\log(\tan\ \phi_{\wedge_j}-\epsilon)}{(\log 3|\eta_0|)^2}+o\left(\frac{1}{(\log 3|\eta_0|)^2}\right). \tag{5.31}$$

From (5.30), (5.31) and (5.28), we get

$$E(f(\eta_1)) \leq \frac{C_6}{|\eta_0|}+\frac{1}{\log 3|\eta_0|}-\frac{\delta}{(\log 3|\eta_0|)^2}+o(\frac{1}{(\log 3|\eta_0|)^2}).$$

Transience is proved. ■

It is worth noting that theorems 5.1.2 and 5.1.5 are in fact contained in theorem 4.1.3.

5.7 Proof of the recurrence

Here for simplicity we also assume that the associated chain has a single essential class with at least two essential states. Let

$$L = L(\mathfrak{A}) < 0. \tag{5.32}$$

We shall show that $\mathcal{L}_T$ is recurrent. As in the proof of transience for any one-dimensional chain $\wedge^{(1)}$ and any $\alpha \in \wedge^{(1)}$, $\alpha \neq 0$, we introduce a Markov process $\zeta_n(\alpha)$, $n \in \mathbf{Z}_+$ (see the definition of this process in section 5.6). To prove the recurrence of $\mathcal{L}_T$ it is sufficient to prove the recurrence of $\zeta_n(\alpha)$.
Let us choose $\tilde{\epsilon} > 0$, $\tilde{\delta} > 0$, $\tilde{n} \in \mathbf{Z}_+$, so that, for any state $\wedge_0^{(2)} \in \mathfrak{A}$, the following inequality holds:

$$\sum_{k=1}^{\tilde{n}}\sum_{\wedge^{(2)}\in\mathfrak{A}} p^{(k)}(\wedge_0^{(2)},\wedge^{(2)})\log(\tan\ \alpha_{\wedge^{(2)}}+\tilde{\epsilon}) < -\tilde{\delta}. \tag{5.33}$$

By (5.32) this is always possible as for any states $\wedge_0^{(2)}$, $\wedge^{(2)} \in \mathfrak{A}$ we have

$$\frac{1}{n}\sum_{k=1}^{n} p^{(k)}(\wedge_0^{(2)},\wedge^{(2)}) \to \pi(\wedge^{(2)}).$$

To prove the recurrence of the chain $\zeta_k(\alpha)$, it is sufficient to prove the recurrence of the chain $\eta_k(\alpha) = \zeta_{\tilde{n}k}(\alpha)$.

To that end, we use the well known recurrence criterion given in chapter 2, theorem 2.2.1, which we recall now:
A Markov chain with the state space $\mathbf{Z}_+$ *and transition probabilities*

p_{ij}, $i,j \in \mathbf{Z}_+$, is recurrent if there exist a non-negative function f on $\mathbf{Z}_+$ and a finite set $A \subset \mathbf{Z}_+$, such that, for any $i \in \mathbf{Z}_+ - A$,

$$\sum_{j \in \mathbf{Z}_+} p_{ij}\, f_j - f_i \leq 0$$

and $f_j \to \infty$ as $j \to \infty$.

We define the function on the state space of η_k putting for any $\alpha \neq 0$

$$f(\alpha) = \log 3|\alpha| \text{ and } f(0) = 0.$$

Let us show that there exists $\mathcal{D} > 0$ such that for any α belonging to the state space of η_k and such that $|\alpha| > \mathcal{D}$, the following inequality holds:

$$E(f(\zeta_{\bar{n}}(\alpha))) \leq f(\alpha). \tag{5.34}$$

After this the proof of recurrence of η_k and so ξ_t will be finished.

Lemma 5.7.1 *Let $\Lambda^{(1)}$ be some one-dimensional non ergodic face and let $\beta \in \Lambda^{(1)} \cap T$. Let $\tau_1(\beta)$ be the first time the process $\xi_t(\beta)$ hits a zero- or one-dimensional face different from $\Lambda^{(1)}$. Then one can find constants $q_1 > 0$, $C_1 > 0$, $\kappa_1 > 0$, which do not depend on β and are such that, for any $t > q_1|\beta|$,*

$$P(\tau_1(\beta) > t) \leq C_1\, e^{-\kappa_1 t}$$

Proof It easily follows from theorem 2.1.7 that, for any $t_1 \in \mathbf{Z}_+$,

$$\begin{aligned} P\left(\tau_1(\beta) > t_1,\ \exists\, t \in \mathbf{Z}_+, t_1 \leq t < \tau_1(\beta), \rho(\xi_t(\beta), \Lambda^{(1)}) < \delta' t_1\right) & \\ \leq c' e^{-\alpha' t_1}, & \end{aligned} \tag{5.35}$$

where $\delta' > 0$, $c' > 0$, $\kappa' > 0$ do not depend on $\beta \in \Lambda^{(1)}$ nor on $t_1 \in \mathbf{Z}_+$. Moreover, due to boundedness of the jumps for $\xi_t(\beta)$, we have, for any $t_1 \in \mathbf{Z}_+$,

$$|\xi_{t_1}(\beta)| \leq |\beta| + dt_1. \tag{5.36}$$

Let us note that, for any two-dimensional chain $\Lambda^{(2)}$, the vector $M_{\Lambda^{(2)}}$ has at least one negative component and therefore, by again using theorem 2.1.7, for any $\Lambda^{(2)}$, any $\alpha \in \Lambda^{(2)} \cap T$ and any $t_2 \in \mathbf{Z}_+$,

$$P(\xi_t(\alpha) \in \Lambda^{(2)},\ t = 1, \ldots, t_2) \leq c''\, e^{h|\alpha| - \kappa'' t_2}, \tag{5.37}$$

where the constants $c' > 0$, $\kappa'' > 0$, $h > 0$ do not depend on $\Lambda^{(2)}, \alpha \in \Lambda^{(2)}$, nor on $t_2 \in \mathbf{Z}_+$. Let now some $t \in \mathbf{Z}_+$ be given. Let

$$t_1 = [\gamma t]\ ,\ t_2 = t - t_1.$$

Then it easily follows from (5.35)–(5.37) that

$$P(\tau_1(\beta) > t) \le c' \, e^{-\alpha'[\gamma t]} + c'' \exp\{h|\beta| + hd\gamma t - \kappa'' t(1-\gamma)\}.$$

Lemma 5.7.1 is now obtained by taking $\gamma > 0$ sufficiently small. ■

Lemma 5.7.2 *Let us choose $\Lambda_1^{(1)}, \Lambda_2^{(1)}$ and*

$$\Lambda_1^{(2)} \in S_+(\Lambda_1^{(1)}), \quad \Lambda_2^{(2)} \in S_-(\Lambda_1^{(1)}) \cap S_+(\Lambda_2^{(1)}).$$

Then there exist constants $q_2 > 0$, $\kappa_2 > 0$, $c_2 > 0$ such that, for any $\beta \in \Lambda_1^{(1)} \cap T$ and any $r > q_2$,

$$P\left(\zeta_1(\beta) \in \Lambda_2^{(2)}, |\zeta_1(\beta)| \ge |\beta|(\tan\phi_{\Lambda_2^{(2)}} + r)\right) \le c_2 \, e^{-\kappa_2 r|\beta|}.$$

Proof Let $\beta \in \Lambda_1^{(1)} \cap T$. Let $\tau_1(\beta)$ be the first time the process $\xi_t(\beta)$ hits some one-dimensional face different from $\Lambda_1^{(1)}$ (we recall that $\zeta_1(\beta) = \xi_{\tau_1(\beta)}(\beta)$).
By lemma 5.7.1, for any $t > q_1|\beta|, t \in \mathbf{Z}_+$, we have

$$P(\tau_1(\beta) > t) \le c_1 \, e^{-\alpha_1 t}. \tag{5.38}$$

On the other hand, from the boundedness of the jumps of $\xi_t(\beta)$,

$$P\left(\xi_{\tau_1(\beta)}(\beta) \in \Lambda_2^{(1)}, |\xi_{\tau_1(\beta)}(\beta)| - |\beta|\tan\phi_{\Lambda_2^{(2)}} > 2d\tau_1(\beta)\right) = 0. \tag{5.39}$$

From (5.38), (5.39), for any $t > q_1|\beta|$, we get

$$P\left(\xi_{\tau_1(\beta)}(\beta) \in \Lambda_2^{(1)}, |\xi_{\tau_1(\beta)}(\beta)| > |\beta|\tan\phi_{\Lambda_2^{(2)}} + 2dt\right) \le c_1 e^{-\alpha t}.$$

This concludes the proof of lemma 5.7.2. ■

Now we shall prove the inequality (5.34). Choose some $\Lambda^{(1)}, \alpha \in \Lambda^{(1)} \cap T$ and $\Lambda_0^{(2)} \in \mathfrak{A}, \Lambda_0^{(2)} \in S_+(\Lambda^{(1)})$. Let $\tilde{n} \in \mathbf{Z}_+$, $\tilde{\epsilon} > 0$, $\tilde{\delta} > 0$ be the constants of (5.33). It follows from lemma 5.6.1 that there exists a constant $\tilde{c} > 0$ such that, for any sequence of two-dimensional faces $\Lambda_1, \ldots, \Lambda_{\tilde{n}} \in \mathfrak{A}$, where $\alpha \in \overline{\Lambda}_0 \cap \overline{\Lambda}_1$, the following inequality holds:

$$|U - V| \le \frac{\tilde{c}}{|\alpha|}, \tag{5.40}$$

where we have set

$$U = P\left(\zeta_k(\alpha) \in \overline{\Lambda}_k \cap \overline{\Lambda}_{k+1}, k=1,\ldots,\tilde{n}, |\zeta_{\tilde{n}}(\alpha)| \le |\alpha| \prod_{j=1}^{\tilde{n}} (\tan\phi_{\Lambda_j} + \tilde{\epsilon})\right),$$

$$V = p(\Lambda_0, \Lambda_1) \cdots p(\Lambda_{\tilde{n}-1}, \Lambda_{\tilde{n}}).$$

Moreover lemma 5.7.2 shows that there exist constants $c>0, \kappa>0, q>0$ such that, for any sequence of two-dimensional faces $\Lambda_1, \ldots, \Lambda_{\tilde{n}} \in \mathfrak{A}$ such that $\alpha \in \overline{\Lambda}_0 \cap \overline{\Lambda}_1$, and for any $k > q$,

$$P\left(\zeta_k(\alpha) \in \overline{\Lambda}_k \cap \overline{\Lambda}_{k+1}, k = 1, \ldots, \tilde{n}, |\zeta_{\tilde{n}}(\alpha)| \geq |\alpha| \prod_{j=1}^{\tilde{n}} (\tan \phi_{\Lambda_j} + k)\right)$$
$$\leq ce^{-\kappa k|\alpha|}. \tag{5.41}$$

From (5.40) and (5.41) it follows that

$$E(f(\zeta_{\tilde{n}}(\alpha))) \leq f(\alpha) + \sum_{\Lambda_1, \ldots, \Lambda_{\tilde{n}} \in \mathfrak{A}} p(\Lambda_0, \Lambda_1) \ldots p(\Lambda_{\tilde{n}-1}, \Lambda_{\tilde{n}})$$
$$\times \log \prod_{j=1}^{\tilde{n}} (\tan \phi_{\Lambda_j} + \tilde{\epsilon}) + \theta(\alpha) ,$$

where $\theta(\alpha) \to 0$ when $|\alpha| \to \infty$.
From the latter inequality and using (5.33) one gets (5.34).
Recurrence is proved. ∎

5.8 Proof of the non-ergodicity

Let there exist an irreducible class of essential states $\mathfrak{A}_i$ for which

$$M(\mathfrak{A}_i) > 0. \tag{5.42}$$

As in the proof of transience, we consider, for any one-dimensional face $\Lambda^{(1)}$ and any $\alpha \in \Lambda^{(1)} \cap T$, the random walk $\xi_t(\alpha)$ with $\xi_0(\alpha) = \alpha$, corresponding to $\mathcal{L}_T$. Define the sequence of stopping times

$$\tau_0(\alpha) = 0, \tau_1(\alpha), \ldots, \tau_n(\alpha), \ldots,$$

where, for $\xi_{\tau_n}(\alpha) \neq 0$, $\tau_{n+1}(\alpha)$ is defined as the next time (after $\tau_n(\alpha)$) the process $\xi_t(\alpha)$ hits either 0 or some one-dimensional face different from that of $\xi_{\tau_n}(\alpha)$. If $\xi_{\tau_n}(\alpha) = 0$, we put $\tau_{n+1}(\alpha) = \tau_n(\alpha)$, etc. Let

$$\begin{aligned}
\zeta_n(\alpha) &= \xi_{\tau_n(\alpha)}(\alpha), \; n \in \mathbf{Z}_+ , \\
\nu(\alpha) &= \inf\{n \in \mathbf{Z}_+ : \xi_{\tau_n(\alpha)} = 0\} .
\end{aligned}$$

From the definition of $\tau_n(\alpha)$, $n \in \mathbf{Z}_+$, it evidently follows that $\tau_{\nu(\alpha)}(\alpha)$ is the first time when $\xi_t(\alpha)$ hits zero, so that $\tau_{\nu(\alpha)+n}(\alpha) = \tau_{\nu(\alpha)}(\alpha), n \in \mathbf{Z}_+$. In order to prove the non-ergodicity of $\xi_t(\alpha)$ it is sufficient to show that

$$E[\tau_n(\alpha)] \to \infty \text{ , and } n \to \infty .$$

Lemma 5.8.1 *For any one-dimensional face $\wedge^{(1)}$, any $\alpha \in \wedge^{(1)}$ and any $n \in \mathbf{Z}_+$,*

$$\tau_{n+1}(\alpha) - \tau_n(\alpha) \geq |\xi_{\tau_n(\alpha)}| \text{ a.s.}$$

To prove this lemma it suffices to note that, for each one-dimensional face $\wedge^{(1)}$ and any $\alpha \in \wedge^{(1)}$,

$$\tau_1(\alpha) \geq |\alpha| \text{ a.s. },$$

and then to use the relation

$$\tau_{n+1}(\alpha) = \tau_n(\alpha) + \tau_1(\xi_{\tau_n(\alpha)}(\alpha)) \text{ a.s.}$$

■

Now let us denote by $\Theta(\mathfrak{A}_i)$ the set of all one-dimensional faces $\wedge^{(1)}$ for which

$$S_-(\wedge^{(1)}) \cap \mathfrak{A}_i \neq \emptyset .$$

We define the following function f on the set of states of $\zeta_n(\alpha) = \xi_{\tau_n(\alpha)}(\alpha)$:

$$f(\beta) = \begin{cases} |\beta|, & \text{if } \beta \in \wedge^{(1)} \cap T \text{ , with } \wedge^{(1)} \in \Theta(\mathfrak{A}_i), \\ 0, & \text{otherwise.} \end{cases}$$

Lemma 5.8.2 *There exist constants $\tilde{n} \in \mathbf{Z}_+, \tilde{\gamma} > 1, \tilde{c} > 0$ such that, for any one-dimensional face $\wedge^{(1)}$ and any $\alpha \in \wedge^{(1)} \cap T$,*

$$E(f(\zeta_{\tilde{n}}(\alpha))) \geq \tilde{\gamma} f(\alpha) - \tilde{c} . \tag{5.43}$$

Proof For any $\wedge^{(1)} \overline{\in} \Theta(\mathfrak{A}_i)$ and any $x \in \overline{\wedge^{(1)}} \cap T$ the inequality (5.43) evidently holds for all $\tilde{\gamma} > 1, \tilde{c} > 0$, $\tilde{n} \in \mathbf{Z}_+$.
Now, consider the case $\wedge^{(1)} \in \Theta(\mathfrak{A}_i)$ and choose $\tilde{n} \in \mathbf{Z}_+$, $\tilde{\epsilon} > 0$ and $\tilde{\gamma} > 1$, so that, for any $\wedge_0^{(2)} \in \mathfrak{A}_i$,

$$\sum_{\wedge_1^{(2)}, \ldots, \wedge_{\tilde{n}}^{(2)} \in \mathfrak{A}_i} p(\wedge_0^{(2)}, \wedge_1^{(2)}) \cdots p(\wedge_{\tilde{n}-1}^{(2)}, \wedge_{\tilde{n}}^{(2)}) \left[\prod_{j=1}^{\tilde{n}} (\tan\, \phi_{\wedge_j^{(2)}} - \tilde{\epsilon}) \right] > \tilde{\gamma}. \tag{5.44}$$

This choice is possible by (5.42) and lemma 5.3.3. Consider now some $\wedge^{(1)} \in \Theta(\mathfrak{A}_i)$, $\wedge_0^{(2)} \in S_+(\wedge^{(1)})$, $\alpha \in \wedge^{(1)} \cap T$. From lemma 5.6.1, it easily follows that

$$E(f(\zeta_{\tilde{n}}(\alpha)))$$

$$\geq |\alpha| \sum_{\Lambda_1^{(2)},\ldots,\Lambda_{\tilde{n}}^{(2)} \in \mathfrak{A}_i} p(\Lambda_0^{(2)}, \Lambda_1^{(2)}) \cdots p(\Lambda_{\tilde{n}-1}^{(2)}, \Lambda_{\tilde{n}}^{(2)}) \left[\prod_{j=1}^{\tilde{n}} (\tan \phi_{\Lambda_j^{(2)}} - \tilde{\epsilon}) \right] - \tilde{c}, \tag{5.45}$$

where $\tilde{c} = c(\tilde{n}, \tilde{\epsilon})$ is a positive constant which does not depend on $\Lambda^{(1)}$ or on $\alpha \in \Lambda^{(1)}$. Then (5.44) follows from (5.45), (5.46).
Lemma 5.8.2 is proved. ∎

Let us now prove (5.43). From lemma 5.8.1 it follows that for any one-dimensional face $\Lambda^{(1)}$, any $\beta \in \Lambda^{(1)} \cap T$ and any $n \in \mathbf{Z}_+$,

$$E(\tau_{n+1}(\beta) - \tau_n(\beta)) \geq E(f(\zeta_n(\beta))). \tag{5.46}$$

From lemma 5.8.2 it follows that, for any $\Lambda^{(1)}$, any $\beta \in \Lambda^{(1)} \cap T$ and $k \in \mathbf{Z}_+$,

$$E(f(\zeta_{k\tilde{n}}(\beta))) \geq \tilde{\gamma}^k \left[f(\beta) - \frac{\tilde{c}}{\tilde{\gamma} - 1} \right], \tag{5.47}$$

where $\tilde{n} \in \mathbf{Z}_+, \tilde{\gamma} > 1, \tilde{c} > 0$ are the constants we sw earlier in lemma 5.8.2.
Let $\Lambda^{(1)} \in \theta_1(\mathfrak{A}_i)$, $\beta \in \Lambda^{(1)} \cap T$ and $f(\beta) = |\beta| > \frac{\tilde{c}}{\tilde{\gamma}-1} + 1$.
Then using (5.47) and (5.48), for any $k \in \mathbf{Z}_+$, we have

$$\begin{aligned} E(\tau_{k\tilde{n}}(\beta)) &\geq \sum_{j=0}^{k} E(\tau_{j\tilde{n}+1}(\beta) - \tau_{j\tilde{n}}(\beta)) \\ &\geq \sum_{j=0}^{k} E(f(\zeta_{j\tilde{n}}(\beta))) \geq \sum_{j=0}^{k} \tilde{\gamma}^k , \end{aligned} \tag{5.48}$$

so that

$$E(\tau_n(\beta)) \to \infty, \text{ as } n \to \infty .$$

The non-ergodicity is proved. ∎

5.9 Queueing applications

First we shall describe two models which in some sense are dual.

Model 1 A single service station has two infinite buffers 1 and 2, $\xi_i(t)$, $i = 1, 2$ being the number of customers in buffer i at time t. There are two independent Poisson arrival streams with intensities λ_i, $i = 1, 2$, to the buffers $i = 1, 2$ respectively.

The station can serve the buffers according to several regimes numbered from 1 to N. For a given regime j, the station serves the buffers, according to independent exponential service times with rates μ_{1j} and μ_{2j} respectively, provided that both buffers are not empty, i.e. $\xi_1(t) \neq 0$, $\xi_2(t) \neq 0$. When one of the buffers is empty, the service rate is taken to be equal to μ. One can imagine changing algorithms, prophylaxis or commuting context, etc.
Let $r(t)$ denote the regime of the station at time t. Let us assume that $r(t) = k$, $\xi_1(t) = 0$, $\xi_2(t) \neq 0$ and let $\tau > t$ be the first arrival of a customer at the buffer i. Our assumption is that, at time τ, *the regime suddenly changes to* $r(\tau) = l$, *with probability* p^1_{kl}, $\sum_{l=1}^N p^1_{kl} = 1$. Similarly, we define the probabilities p^2_{kl}.
If$\xi_1(t) = \xi_2(t) = 0$, then we do not specify the protocol, since it has no influence on the ergodicity of the system.

Model 2 Let us consider a customer (computer, bus etc.) which has access to N databases, but can use at time t at most two of them, $r_1(t)$ and $r_2(t)$, where $r_i(t) \in \{1, \ldots, N\}$, $i = 1, 2$. This user has two buffers 1,2 accumulating data coming from the databases $r_1(t)$ and $r_2(t)$, respectively. Let λ_i be the intensity of arrivals from the database i, $i = 1, 2$. Let $\xi_i(t)$ be the number of units in buffer i at time t (necessarily issuing from the database $r_i(t)$).

Under the condition $\xi_1(t) > 0$, $\xi_2(t) > 0$ and $r_i(t) = k_i$, $i = 1, 2$, the customer *serves* the buffers with the following intensities:

(a) $\mu^i_{k_i}$, when serving one unit from buffer i;
(b) $\mu_{k_1 k_2}$, when, simultaneously, one unit from buffer 1 and one unit from buffer 2 are served.

When e.g. $\xi_1(t) = 0$, $\xi_2(t) \neq 0$, the user switches to another database. The intensity of switching to database k is

$$\lambda_1(k; l, \xi_2(t)) ,$$

where l is the later database used by buffer 1. In a similar way, we introduce

$$\lambda_2(k; \xi_1(t), l) .$$

It is also assumed that the above swiching is accompanied with the arrival of one unit from the database K at the corresponding buffer.

In model 1, a *regime* corresponds to a quarter-plane. In model 2, a single database corresponds to a one-dimensional complex and a pair of databases corresponds to a quarter-plane. Both models can be treated with similar methods, since they induce random walks in two-dimensional complexes. They do not completely fit our main theorems, as the probabilities of jumps from one-dimensional faces depend on the last visited quarter-plane. We could solve this case also, by applying similar methods. Instead of this, we shall discuss in more detail a particular case of model 1.

Analysis of model 1

Let us consider the set P of all $2N$ pairs (i,k), $i=1,2$ and $k=1,\dots,N$. Thus $(1,k)$ and $(2,k)$ can be considered as the one-dimensional faces of the k-th quarter-plane. We assume that P is subdivided into $M \le 2N$ equivalence classes α, so that (i,k) and (j,l) belong to different classes if $i \ne j$. This is equivalent to a *gluing* of the corresponding one-dimensional faces. Let $\alpha(i,k)$ be the class to which (i,k) belongs.
So we assume that $p^i_{kl}=0$, if (i,k) and (i,l) belong to different classes and that p^i_{kl} depends only on $\alpha = \alpha(i,k)$. We shall write *ad libitum* $p_{\alpha,l}$ instead of p^i_{kl} if $(i,k)\in\alpha$. Our process is a triple $(\xi_1(t),\ \xi_2(t),\ r(t))$ and forms a countable Markov chain.

Correspondence with the random walk

Let us consider the N quarter-planes $(\mathbf{Z}^2_+)_i$, $i=1,\dots,N$, glued together and let $r(t)$ take the values $1,\dots,N$. The one-dimensional faces correspond to the equivalence classes which have been introduced in the above gluing. This defines in fact a complex T. It will be convenient to consider an embedded discrete time Markov chain instead of the original chain which is in continuous time. To that end, we choose $c>0$, sufficiently large (the ergodicity conditions do not depend on the choice of this constant), and we define the following transition probabilities on T.

For $k,l\ge 1, (k,l)_i \in (\mathbf{Z}^2_+)_i$, we can write

$$cp[(k,l)_i,(k',l')_i] = \begin{cases} \lambda_1\ , \ k'=k+1,\ l'=l\ , \\ \lambda_2\ , \ k'=k,\ l'=l+1\ , \\ \mu_{1i}\ , \ k'=k-1,\ l'=l\ , \\ \mu_{2i}\ , \ k'=k,\ l'=l-1\ . \end{cases}$$

On the boundaries,

$$
\begin{aligned}
cp[(0,l)_i,\ (0,l')_i] &= \begin{cases} \lambda_2 & , \quad l' = l+1\,, \\ \mu & , \quad l' = l-1\,, \end{cases} \\
cp[(0,l)_\alpha,\ (1,l')_i] &= \lambda_1 p_{\alpha i}\ , \quad l' = l\,, \\
cp[(k,0)_j,\ (k',0)_j] &= \begin{cases} \lambda_1 & , \quad k' = k+1\,, \\ \mu & , \quad k' = k-1\,, \end{cases} \\
cp[(k,0)_\alpha, (k',1)_j] &= \lambda_2 p_{\alpha j}\ , \quad k' = k\,.
\end{aligned}
$$

We write $(0,l)_\alpha = (0,l)_i$, if $(i,l) \in \alpha$ and, similarly, $(k,0)_\alpha = (k,0)_j$ if $(j,k) \in \alpha$. The probabilities $p[(k,l)_i,\ (k,l)_i]$ are defined by the normalization condition.

Mean jumps Up to the multiplicative constant C^{-1}, we have the following possibilities:

- if $\wedge_i^{(2)}$ corresponds to the *regime* i, then

$$M_i \stackrel{\text{def}}{=} M_{\wedge_i^{(2)}} = (\lambda_1 - \mu_{1i},\ \lambda_2 - \mu_{2i})\ ;$$

- if $\wedge_\alpha^{(1)}$ corresponds to the equivalence class α, then

$$M_{\alpha\alpha} \stackrel{\text{def}}{=} Pr_{\wedge_\alpha^{(1)}}\,[M_{\wedge_\alpha^{(1)},\wedge_\alpha^{(2)}}] = \lambda_\alpha - \mu,$$

  where $\lambda_\alpha = \lambda_1$ (resp. λ_2), if $\xi_2(t) = 0$ (resp. $\xi_1(t) = 0$) ;

$$M_{\alpha,l} \stackrel{\text{def}}{=} M_{\wedge_\alpha^{(1)},\wedge_l^{(2)}} = \begin{cases} (0, \lambda_2 p_{\alpha l}), & \text{if } \alpha \text{ corresponds to } \xi_2(t) = 0, \\ (\lambda_1 p_{\alpha l}, 0), & \text{if } \alpha \text{ corresponds to } \xi_1(t) = 0. \end{cases}$$

(Recall that $\xi_i(t)$ is the content of buffer i, $i = 1, 2$, at time t.)

Transience and ergodicity Should there exist only one regime, e.g. j, then necessary and sufficient conditions for the system to be ergodic for all $\mu > 0$ sufficiently large would be

$$\lambda_i < \mu_{ij}\ , \quad i = 1, 2\,. \tag{5.49}$$

In fact, for several regimes, sufficient ergodicity conditions are: *inequalities (5.49) hold for all* $j = 1, \ldots, N$ *and* μ *sufficiently large.* In this case, the system is ergodic, independently of the p_{kl}^i's. As is to be expected, these conditions are not necessary and we get, below, the exact necessary and sufficient conditions, which connect the parameters p_{kl}^i, λ_i, μ_{ij} together. In particular, we show that if, for some i, j, we have

$$\lambda_i > \mu_{ij}\ ,$$

then, for some p_{kl}^i, the system may be ergodic. However, theorem 5.1.2 yields the following.

Corollary 5.9.1 *If there exists j such that*

$$\lambda_1 > \mu_{1j}\ ,\ \lambda_2 > \mu_{2j}\ , \tag{5.50}$$

then the system is transient. ∎

We note that a one-dimensional face $\Lambda_\alpha^{(1)}$ (or simply the equivalence class α) such that $(1,k) \in \alpha$ is ergodic if and only if $\lambda_2 < \mu_{2k}$. In this case, we have (see section 5.1).

$$\operatorname{sgn} V_{\Lambda_\alpha^{(1)}} = \operatorname{sgn}\left[\lambda_2 - \mu + \sum_{k_i(1,k)\in\alpha}\left[(\lambda_1 - \mu_{1k})\frac{\lambda_2 p_{\alpha k}}{\lambda_2 - \mu_{2k}}\right]\right].$$

One could write similar formulas for ergodic faces $\Lambda_\alpha^{(1)}$, such that $(2,j) \in \alpha$. This will be summarized now. Let us assume that, for all ergodic α,

$$\lambda_\alpha - \mu + \sum_{ki:(i,k)\in\alpha}\frac{\lambda_i - \mu_{ik}}{\lambda_j - \mu_{jk}} \neq 0.$$

Then the following proposition holds.

Corollary 5.9.2 *For the Markov chain $\mathcal{L}_T$ to be ergodic, it is necessary that, for all ergodic α,*

$$\lambda_\alpha - \mu + \sum_{k:(i,k)\in\alpha}\frac{\lambda_i - \mu_{ik}}{\lambda_j - \mu_{jk}}\lambda_j p_{\alpha k} < 0\ . \tag{5.51}$$

Theorem 5.3.2 gives the necessary and sufficient conditions for ergodicity in the cases which are not covered by corollaries 5.9.1 and 5.9.2. To show even more explicit computations let us consider a system admitting only two classes, α_1 and α_2. In the usual language of queueing theory, this means that the quantities

$$p_{kl}^1 = p_l^1\ ,\ p_{kl}^2 = p_l^2 \tag{5.52}$$

do not depend on k. Let us classify the regimes in the latter case. The regime j will be said to be 1-*outgoing* if

$$\lambda_2 > \mu_{2j}\ . \tag{5.53}$$

We have seen in this case that, for the system to be ergodic, it is necessary to have

$$\lambda_1 < \mu_{1j}\ .$$

The set of all 1-*outgoing* regimes will be denoted by A_1. Similarly, we define 2-*outgoing* regimes A_2, when

$$\lambda_2 < \mu_{2j}\ ,\ \lambda_1 > \mu_{1j}\ . \tag{5.54}$$

Clearly, for all other regimes (belonging to $\{1,\ldots,n\}\backslash(A_1 \cup A_2)$), we have

$$\lambda_1 < \mu_{ij},\ \lambda_2 < \mu_{2j}\ .$$

Let us note that the stationary probabilities of the associated Markov chain (irreducible and aperiodic) are given by

$$\pi_i \stackrel{\text{def}}{=} \pi(\wedge_i^{(2)}) = \frac{1}{2} p_{sc}(\wedge^{(1)}, \wedge_i^{(2)}), \tag{5.55}$$

where $\wedge^{(1)}$ is α_1 for 1-*outgoing* $\wedge_i^{(2)}$ and α_2 for 2-*outgoing* $\wedge_i^{(2)}$. Using (5.2), we have, e.g. for 2-*outgoing* $\wedge^{(2)}$,

$$p_{sc}(\wedge^{(1)}, \wedge^{(2)}) = \frac{q_{01}^{\wedge^{(2)}} \gamma_{\wedge^{(2)}}}{\sum_{\wedge^{(2)}} q_{01}^{\wedge^{(2)}} \gamma_{\wedge^{(2)}}},$$

where $\gamma = \gamma_{\wedge_i^{(2)}}$ is defined from equation (5.4), which reduces to

$$\lambda_i(1-\gamma) + \mu_{1i}(1-\gamma)^{-1} - (\lambda_i + \mu_{1i}) = 0\ ,$$

whence

$$\gamma = 1 - \frac{\mu_{1i}}{\lambda_i}\ .$$

Define the following real number:

$$M_1 = \frac{\sum_{j\in A_1} \log\left[\frac{|\lambda_2 - \mu_{2j}|}{|\lambda_1 - \mu_{1j}|}\right] \lambda_2 p_j^2 \left(1 - \frac{\mu_{2j}}{\lambda_2}\right)}{\sum_{j\in A_1} \lambda_2 p_j^2 \left(1 - \frac{\mu_{2j}}{\lambda_2}\right)}\ .$$

We can define M_2 in a similar way. Our final result, for the example under consideration, is contained in the next

Theorem 5.9.3 *Assume in model 1 that (5.52) holds and both A_1 and A_2 are not empty. Then the system is recurrent if*

$$M_1 + M_2 < 0$$

and transient if

$$M_1 + M_2 > 0\ .$$

Let us consider a system with only two classes α_1 and α_2, under the following assumptions:

- there is only one 1-outgoing regime $j = 1$, and

$$\lambda_2 > \mu_{21} \,,\; \lambda_1 < \mu_{11} \,;$$

- there are only two 2-outgoing regimes $j = 2, 3$,

$$\lambda_2 < \mu_{22} \,,\; \lambda_1 > \mu_{12} \,,$$

and

$$\lambda_2 < \mu_{23} \,,\; \lambda_1 > \mu_{13} \,;$$

- for the other regimes $j = 4, ..., N$, we have

$$\lambda_2 < \mu_{2j} \text{ and } \lambda_1 < \mu_{1j} \,.$$

Then the matrix (5.9) in the present case is given by

$$\begin{aligned}
A_{11} &= A_{22} = A_{33} = A_{23} = A_{32} = 0 \,, \\
A_{21} &= \sqrt{\tan \phi_{\Lambda_2^{(2)}} \tan \phi_{\Lambda_1^{(2)}}} \,, \\
A_{31} &= \sqrt{\tan \phi_{\Lambda_3^{(2)}} \tan \phi_{\Lambda_1^{(2)}}} \,, \\
A_{12} &= \sqrt{\tan \phi_{\Lambda_1^{(2)}} \tan \phi_{\Lambda_2^{(2)}} \, p_2^2} \,, \\
A_{13} &= \sqrt{\tan \phi_{\Lambda_1^{(2)}} \tan \phi_{\Lambda_3^{(2)}} \, p_3^2} \,.
\end{aligned}$$

One can show that the maximal eigenvalue of this matrix is given by

$$\lambda_1^2 = p_2^2 \tan \phi_{\Lambda_1^{(2)}} \tan \phi_{\Lambda_2^{(2)}} + p_3^2 \tan \phi_{\Lambda_1^{(2)}} \tan \phi_{\Lambda_3^{(2)}} \,.$$

The above conditions imply the following result.

Theorem 5.9.4 *Assume in model 1 that (5.52) holds and $A_1 = \{1\}$, $A_2 = \{2, 3\}$. Then the system is ergodic if*

$$M = \frac{1}{2} \log \left[p_2^2 \tan \phi_{\Lambda_1^{(2)}} \tan \phi_{\Lambda_2^{(2)}} + p_3^2 \tan \phi_{\Lambda_1^{(2)}} \tan \phi_{\Lambda_3^{(2)}} \right] < 0$$

and non-ergodic if

$$M > 0 \,.$$

Note that under the conditions of theorem 5.9.4 it is not necessary to calculate the maximal eigenvalue of the matrix (5.9). It is sufficient to note that, for any $n \in \mathbf{Z}_+$ in (5.7), we have

$$E\left(\prod_{j=1}^{2n} \tan \phi_{\Lambda_j} | \Lambda_0 = \Lambda_1^{(2)}\right) = \left[E\left(\tan \phi_{\Lambda_1} \tan \phi_{\Lambda_2} | \Lambda_0 = \Lambda_1^{(2)}\right)\right]^n .$$

5.10 Remarks and problems

Remark 1 In our case, we could have defined an associated Markov chain with a different set of states: $\{\Lambda^{(1)}\}$ instead of $\{\Lambda^{(2)}\}$ with transition probabilities

$$p(\Lambda^{(1)}, \Lambda_1^{(1)}) = p(\Lambda^{(2)}, \Lambda_1^{(2)}) ,$$

for any $\Lambda^{(2)} \in S_+(\Lambda^{(1)})$ and the unique $\Lambda_1^{(2)} \in S_+(\Lambda_1^{(1)}) \cap S_-(\Lambda^{(1)})$, if such $\Lambda_1^{(2)}$ exists, and 0 otherwise. In the more general case when Condition A1 of section 5.1 (boundedness of jumps) is not satisfied, the situation becomes more complicated : the scattering probabilities depend on the ingoing bristle of the hedgehog. Thus the probabilities $p(\Lambda^{(2)}, \Lambda_1^{(2)})$ and $p(\Lambda_2^{(2)}, \Lambda_1^{(2)})$ can be different for different $\Lambda^{(2)}, \Lambda_2^{(2)} \in S_+(\Lambda^{(1)})$. But this case can also be taken care of, using the same therapy.

Remark 2 It is of interest to generalize our results to the case when the simplexes of our complex are not $\mathbf{Z}_+^2$, but angles in $\mathbf{R}_+^2$ or in $\mathbf{Z}_+^{(2)}$. Of special interest is the situation when these angles in $\mathbf{Z}_+^2$ are not commensurable with π.

Remark 3 Our methods permit the classification of random walks in $\mathbf{Z}_+^N$ under the same *non-zero* and homogeneity assumptions, when all vectors of mean jumps inside all faces Λ, with $\dim \Lambda \geq 3$, have their coordinates negative. Then M_Λ, with $\dim \Lambda = 1$ or 2, will become vectors which can be determined from the corresponding induced chains.

Remark 4 We want to show now that all our assumptions have *Lebesgue measure* 1 in the parameter space. Indeed, assumptions $0_1, 0_3$ are fulfilled when all the $p_{\alpha\beta}$'s are positive. Assumption 0_2 is satisfied, except for a finite number of hyperplanes. Finally, assumptions $0_4, 0_5$ are fulfilled except when $v_{\Lambda^{(1)}}$ in lemma 5.1.6 is equal to zero.

6
Stability

The present chapter is devoted to the continuity of stationary distributions for families of homogeneous irreducible and aperiodic Markov chains. In section 6.1 we give a necessary and sufficient condition for this continuity. In section 6.2 we present constructive sufficient conditions for the continuity of the stationary probabilities in terms of *test* functions. Finally, section 6.3 deals with the continuity of random walks in $\mathbf{Z}^N$.

6.1 A necessary and sufficient condition for continuity

Let us consider a family of homogeneous irreducible aperiodic Markov chains $\{L^\nu\}$ with discrete time and countable set of states $\mathcal{A} = \{0, 1, \ldots\}$, for $\nu \in D$, where D is an open subset of the real line. By $p_{ij}(t, \nu)$ we denote the t-step transition probability from the point i to the point j in L^ν. Everywhere in this chapter we assume that the $p_{ij}(1, \nu)$ are continuous in ν for all $\nu \in \mathcal{D}$ and $i, j \in \mathcal{A}$. For the sake of brevity, we will write

$$p_{ij}(\nu) \stackrel{\text{def}}{\equiv} p_{ij}(1, \nu)\,.$$

Lemma 6.1.1 *The $p_{ij}(t, \nu)$ are continuous functions of ν ($\nu \in D$) for every natural number t and all $i, j \in \mathcal{A}$.*

Proof We prove the lemma by induction. For $n = 1$ the function $p_{ij}(1, \nu)$ is continuous in ν for any $i, j \in \mathcal{A}$. We have

$$p_{ij}(n+1, \nu) = \sum_{k=0}^{\infty} p_{ik}(\nu) p_{kj}(n, \nu) = \sum_{k=0}^{\infty} \phi_k(\nu). \tag{6.1}$$

It follows from the inclusion hypothesis that the $\phi_k(\nu)$ are continuous functions of ν for any k. We have

$$\sum_{k=0}^{\infty} p_{ik}(\nu) \equiv 1 \quad (\nu \in D) . \tag{6.2}$$

Consequently, the series (6.2) consisting of continuous functions converges uniformly with respect to $\nu \in D$, and $p_{ik}(n, \nu) \leq 1$ for any k, i, n and $\nu \in D$. These two conditions lead to the uniform convergence of the series (6.1). The sum of a uniformly convergent series of continuous functions is a continuous function. Consequently, the $p_{ij}(n+1, \nu)$ are continuous in ν. The lemma is proved. ■

On the set $\mathcal{A}$ let $\{\pi_j(\nu), j \in \mathcal{A},\ \nu \in D\}$ be a given family of distributions, where D is some open subset of the real line. We have

$$\sum_{j \in \mathcal{A}} \pi_j(\nu) = 1 \quad (\nu \in D) .$$

Definition 6.1.2 *The family of distributions $\{\pi_j(\nu), j \in \mathcal{A},\ \nu \in D\}$ satisfies Condition λ at the point $\nu_0 \in D$ if, for any $\epsilon > 0$, there exist $\delta > 0$ and a finite set $B^\epsilon \subset \mathcal{A}$ such that*

$$\sum_{j \in \mathcal{A} \setminus B^\epsilon} \pi_j(\nu) < \epsilon , \tag{6.3}$$

for all ν with $\mid \nu - \nu_0 \mid < \delta$.

Let the chains L^ν be ergodic for every ν belonging to some neighbourhood $U_0 \subset D$ of zero.

Theorem 6.1.3 *The stationary probabilities $\pi_j(\nu)$ depend on ν continuously at $\nu = 0$ for all $j \in \mathcal{A}$ if, and only if, the family of distributions $\{\pi_j(\nu)\}$ satisfies Condition λ at the point $\nu = 0$.*

Before proving this, we make the following remark. Following Prohorov [Pro56], we form the metric space $D(\mathcal{A})$. To that end, we define the distance $L(\mu_1, \mu_2)$ between any two measures μ_1 and μ_2 on $\mathcal{A} = \{0, 1, \ldots, n\}$, so that convergence in the sense of this distance is equivalent to weak convergence of measures. The collection of all measures on $\mathcal{A}$ together with the function $L(\mu_1, \mu_2)$ forms the metric space $D(\mathcal{A})$. Still in accordance with [Pro56] we introduce the following definition.

Definition 6.1.4 *A set T of measures on $\mathcal{A}$ satisfies Condition χ if*

(χ_1) *the values $\mu(\mathcal{A}), \mu \in T$, are bounded,*

(χ_2) *for any given $\epsilon > 0$, there exists a finite set K_ϵ of points such that $\mu(\mathcal{A} \setminus K_\epsilon) < \epsilon$, for all $\mu \in T$.*

In [Pro56] it is proved that for $t \subset D(\mathcal{A})$ to be compact, it is necessary and sufficient that Condition χ be satisfied. For $\{\pi_j(\nu)\}$, Condition χ obviously implies Condition λ. Therefore, as a result of theorem 6.1.3, we have the following theorem.

Theorem 6.1.5 *In order that the stationary probabilities $\pi_j(\nu)$ depend continuously on ν for all $j \in \mathcal{A}$ it is sufficient that the family $\{\pi_j(\nu)\}$ of distributions be compact in $D(\mathcal{A})$.*

Let us pass to the proof of theorem 6.1.3. Let $\{\pi_j(\nu)\}$ satisfy Condition λ at $\nu = 0$. We prove that $\pi_j(\nu)$ is right-continuous at $\nu = 0$ for any $j \in \mathcal{A}$ (the left continuity can be proved analogously). Take an arbitrary $\epsilon > 0$. Condition λ implies the existence of a $\nu_0 > 0$ and a finite set $k^\epsilon \subset \mathcal{A}$ such that for any ν for which

$$0 \leq \nu \leq \nu_0 \qquad \text{(Condition } a_1\text{)}$$

we have the inequality

$$\sum_{k \in \mathcal{A} \setminus B^\epsilon} \pi_k(\nu) < \epsilon \,. \tag{6.4}$$

We prove that, for any $j \in \mathcal{A}$, there is a $\nu_1(j)$ such that, for $0 < \nu < \nu_1(j)$,

$$|\pi_j(\nu) - \pi_j(0)| < 10\epsilon \,. \tag{6.5}$$

By the same token, we prove the continuity of the $\pi_j(\nu)$ at $\nu = 0$. The following inequality is satisfied for any t, i and j:

$$\begin{aligned} |\pi_j \quad (\nu) \quad & -\pi_j(0)| \\ = \quad & |p_{ij}(t,\nu) - p_{ij}(t,0) + \pi_j(\nu) - p_{ij}(t,\nu) + p_{ij}(t,0) - \pi_j(0)| \\ \leq \quad & |p_{ij}(t,\nu) - p_{ij}(t,0)| + |\pi_j(\nu) - p_{ij}(t,\nu)| + |p_{ij}(t,0) - \pi_j(0)| \,. \end{aligned} \tag{6.6}$$

From the ergodicity of the chains L^ν and L^0, for fixed i and j, it follows that there is a $t_0(\nu)$ such that, for

$$t > t_0(\nu) \qquad \text{(Condition } a_2\text{)} \,,$$

we have

$$
\begin{aligned}
|\pi_j(\nu) - p_{ij}(t,\nu)| &< \epsilon\,, \\
|p_{ij}(t,0) - \pi_j(0)| &< \epsilon\,.
\end{aligned} \tag{6.7}
$$

Consider the first term on the right side of (6.6). For any $T < t$, we have

$$
\begin{aligned}
&|p_{ij}(t,\nu) - p_{ij}(t,0)| \\
&= \left| \sum_{k\in B^\epsilon} p_{ik}(t-T,\nu)p_{kj}(T,\nu) + \sum_{k\in \mathcal{A}\backslash B^\epsilon} p_{ij}(t-T,\nu)p_{kj}(T,\nu) \right. \\
&\qquad \left. - \sum_{k\in B^\epsilon} p_{ik}(t-T,0)p_{kj}(T,0) - \sum_{k\in \mathcal{A}\backslash B^\epsilon} p_{ik}(t-T,0)p_{ki}(T,0) \right| \\
&\leq \left| \sum_{k\in B^\epsilon} [p_{ik}(t-T,\nu)p_{kj}(T,\nu) - p_{ik}(t-T,0)p_{kj}(T,0)] \right| \\
&\qquad + \sum_{k\in \mathcal{A}\backslash B^\epsilon} p_{ik}(t-T,\nu) + \sum_{k\in \mathcal{A}\backslash B^\epsilon} p_{ik}(t-T,0)\,.
\end{aligned} \tag{6.8}
$$

From (6.4) and the ergodicity of L^ν and L^0 it follows that there is a $t_1(\nu)$ such that, for

$$
t - T > t_1(\nu) \qquad \text{(Condition } a_3\text{)}\,,
$$

we have

$$
\left\{
\begin{aligned}
\textstyle\sum_{k\in \mathcal{A}\backslash B^\epsilon} p_{ik}(t-T,\nu) &< 2\epsilon\,, \\
\textstyle\sum_{k\in \mathcal{A}\backslash B^\epsilon} p_{ik}(t-T,0) &< 2\epsilon\,.
\end{aligned}
\right. \tag{6.9}
$$

Let us pass to the study of the first term on the right side of (6.8).

$$
\begin{aligned}
&\left| \sum_{k\in B^\epsilon} [p_{ik}(t-T,\nu)p_{kj}(T,\nu) - p_{ik}(t-T,0)p_{kj}(T,0)] \right| \\
&= \left| \sum_{k\in B^\epsilon} [p_{ik}(t-T,\nu)p_{ij}(T,\nu) - p_{ik}(t-T,\nu)[p_{ij}(T,\nu) - p_{kj}(T,\nu)] \right. \\
&\qquad \left. - p_{ik}(t-T,0)p_{ij}(T,0) + p_{ik}(t-T,0)[p_{ij}(T,0) - p_{kj}(T,0)]] \right| \\
&\leq \sum_{k\in B^\epsilon} p_{ik}(t-T,\nu)|p_{ij}(T,\nu) - p_{kj}(T,\nu)| \\
&\qquad + \sum_{k\in B^\epsilon} p_{ik}(t-T,0)|p_{ij}(T,0) - p_{kj}(T,0)| \\
&\qquad + \left| p_{ij}(T,\nu) \sum_{k\in B^\epsilon} p_{ik}(t-T,\nu) - p_{ij}(T,0) \sum_{k\in B^\epsilon} p_{ik}(t-T,0) \right|
\end{aligned}
$$

$$\leq \left| p_{ij}(T,\nu) \sum_{k \in B^\epsilon} p_{ik}(t-T,\nu) - p_{ij}(T,0) \sum_{k \in B^\epsilon} p_{ik}(t-T,0) \right|$$
$$+ \sum_{k \in B^\epsilon} |p_{ij}(T,\nu) - p_{kj}(T,\nu)| + \sum_{k \in B^\epsilon} |p_{ij}(T,0) - p_{kj}(T,0)|. \quad (6.10)$$

It follows from (6.9) that, for $t - T > t_1(\nu)$, we have

$$\begin{cases} 1 - \sum_{k \in B^\epsilon} p_{ik}(t-T,\nu) & < & 2\epsilon\,, \\ 1 - \sum_{k \in B^\epsilon} p_{ik}(t-T,0) & < & 2\epsilon\,. \end{cases} \quad (6.11)$$

Since $p_{ij}(T,\nu)$ is a continuous function of ν at zero for a fixed T, there is a $\nu_2 = \nu_2(T)$ such that, for

$$0 < \nu < \nu_2(T) \quad (\text{Condition } a_4)\,,$$

we have

$$|p_{ij}(T,\nu) - p_{ij}(T,0)| < \epsilon\,. \quad (6.12)$$

It follows from (6.11) and (6.12) that

$$\left| p_{ij}(T,\nu) \sum_{k \in B^\epsilon} p_{ik}(t-T,\nu) - p_{ij}(T,0) \sum_{k \in B^\epsilon} p_{ik}(t-T,0) \right| \leq 3\epsilon\,. \quad (6.13)$$

The chain L^0 is ergodic. Consequently, there exists

$$T = T(B^\epsilon) \quad (\text{Condition } a_5)$$

for which

$$\max_{k \in B^\epsilon} |p_{ij}(T,0) - p_{kj}(T,0)| < \frac{\epsilon}{M}\,, \quad (6.14)$$

$$\sum_{k \in B^\epsilon} |p_{ij}(T,0) - p_{kj}(T,0)| < \epsilon\,, \quad (6.15)$$

where M is equal to the number of terms in the sum (6.15). The continuity of $p_{kj}(T,\nu)$ at $\nu = 0$ for a fixed T implies the existence of $\nu_3(T)$ such that, for

$$\nu < \nu_3(T) \quad (\text{Condition } a_6)\,,$$

we have

$$\sup_{k \in B^\epsilon, \nu < \nu_3(T)} |p_{kj}(T,\nu) - p_{kj}(T,0)| < \frac{\epsilon}{M}\,. \quad (6.16)$$

Inequalities (6.14) and (6.16) yield

$$\sup_{k \in B^\epsilon, \nu < \nu_3(T)} |p_{ij}(T,\nu) - p_{kj}(T,\nu)| < \frac{2\epsilon}{M}\,. \quad (6.17)$$

Consequently,

$$\sum_{k\in B^\epsilon} |p_{ij}(T,\nu) - p_{kj}(T,\nu)| < 2\epsilon \,. \tag{6.18}$$

Comparing (6.9), (6.13), (6.15), and (6.18), we obtain that if t, T and ν satisfy Conditions a_1–a_6, then

$$|\pi_j(\nu) - \pi_j(0)| < 10\epsilon \,. \tag{6.19}$$

The proof of the right continuity of $\pi_j(\nu)$ at $\nu = 0$ will be complete if we show there exist T and $\nu_1 > 0$ such that, for $0 < \nu < \nu_1$, there is a t depending on ν for which Conditions a_1–a_6 are satisfied. This can be shown easily in the following way. For the set B^ϵ we find $T = T(B^\epsilon)$ (Condition a_5), and for T we find numbers $\nu_2(T)$ and $\nu_3(T)$. Set

$$\nu_1(T) = \min\,\{\nu_0,\ \nu_2(T),\ \nu_3(T)\}\,.$$

Then, for any $\nu < \nu_1(T)$, the choice $t > \max\,\{t_0(\nu), T + t_1(\nu)\}$ fulfils Conditions a_1–a_6.

Now we prove that the continuity of $\pi_j(\nu)$ at $\nu = 0$ for any $j \in \mathcal{A}$ implies the satisfaction of Condition λ for $\{\pi_j(\nu)\}$ at $\nu = 0$. The chain L^0 is ergodic. Consequently, for any $\epsilon > 0$ there is a finite set B^ϵ for which

$$\sum_{k\in \mathcal{A}\setminus B^\epsilon} \pi_k(0) < \frac{\epsilon}{2}\,,$$

$$\sum_{k\in B^\epsilon} \pi_k(0) > 1 - \frac{\epsilon}{2}\,.$$

The continuity of $\pi_k(\nu)$ at $\nu = 0$ implies that there exists ν_0 such that, for $|\nu| < v_0$, we have

$$\max_{k\in B^\epsilon} |\pi_k(\nu) - \pi_k(0)| < \frac{\epsilon}{2M}\,, \tag{6.20}$$

where M is the number of elements in the set B^ϵ. Therefore,

$$\sum_{k\in B^\epsilon} |\pi_k(\nu) - \pi_k(0)| < \frac{\epsilon}{2}\,, \tag{6.21}$$

$$\sum_{k\in \mathcal{A}\setminus B^\epsilon} \pi_k(\nu) < \epsilon\,. \tag{6.22}$$

Consequently, $\{\pi_j(\nu)\}$ satisfies Condition λ at $\nu = 0$. The theorem is proved. ■

6.2 Sufficient conditions for the continuity of stationary probabilities

As in section 6.1, let us consider a family $\{L^\nu\}$ of irreducible aperiodic Markov chains, with transition probabilities $p_{ij}(1,\nu) \equiv p_{ij}(\nu)$ continuously depending on ν, for $\nu \in D \subset \mathbf{R}$ (D is an open set).
The theorem of this section will be formulated in terms of *test* functions. Later, the continuity of stationary probabilities of random walks in $\mathbf{Z}_+^n$ will be studied by means of the results of the present section.
Assume that on the set $\mathcal{A} = \{0, 1, \ldots\}$ there are given two families $f^\nu = \{f_i^n\}$ and $g^\nu = \{g_i^\nu\}$, for $i \in \mathcal{A}, \nu \in D$, of real functions, where

$$\inf_{i\in\mathcal{A},\nu\in D} f_i^\nu \geq 0, \qquad \inf_{i\in\mathcal{A},\nu\in D} g_i^\nu = \delta > 0 \,.$$

Theorem 6.2.1 *Assume that for some finite non-empty set $B \subset \mathcal{A}$ the functions $\{f_i^\nu\}$ and $\{g_i^\nu\}$ satisfy the following conditions:*

(i) $\displaystyle\sum_{j=0}^{\infty} p_{ij}(\nu) f_j^\nu - f_i^\nu < -g_i^\nu \quad i \notin B,\ \nu \in D.$

(ii) $\displaystyle\sup_{i\in B,\nu\in D} \sum_{j=0}^{\infty} p_{ij}(\nu) f_j^\nu = \lambda < \infty.$

(iii) $g_i^\nu \to \infty$ *as* $i \to \infty$ *uniformly in* $\nu \in D$.

Then the chains L^ν are ergodic for every $\nu \in D$, and the stationary probabilities $\pi_j(\nu)$ are continuous in ν, for $\nu \in D$ and $j \in \mathcal{A}$.

Proof The ergodicity of L^ν for every $\nu \in D$, follows from the hypotheses of theorem 2.2.3. We define by induction

$$\begin{aligned} y_i^1(\nu) &= y_i(\nu) = f_i^\nu \,, \\ y_j^{n+1}(\nu) &= \sum_{j=0}^{\infty} p_{ij}(\nu) y_j^n(\nu) \,. \end{aligned}$$

It is obvious that $y_i^n(\nu) \geq 0$, for any natural numbers i and n, and any $\nu \in D$. We have

$$\begin{aligned} y_i^2(\nu) &= \sum_{j=0}^{\infty} p_{ij}(\nu) f_j^\nu < f_i^\nu - g_i^\nu = y_i(\nu) - g_i^\nu \,,\ i \notin B \,, \\ y_i^2(\nu) &\leq \lambda,\ i \in B \,, \\ y_i^3(\nu) &= \sum_{j=0}^{\infty} p_{ij}(\nu) y_j^2(\nu) \leq \lambda \sum_{j=B} p_{ij}(\nu) + \sum_{j\notin B} p_{ij}(\nu)[y_i(\nu) - g_i^\nu] \,. \end{aligned}$$

Write

$$\lambda_1 = \sup_{i\in B, v\in D} g_i^\nu; \quad p_{iB}(n,\nu) = \sum_{j\in B} p_{ij}(n,\nu) .$$

After easy calculations we obtain

$$y_i^3(\nu) \le (\lambda+\lambda_1)p_{iB}(\nu) + y_i^2(\nu) - \sum_{j=0}^{\infty} p_{ij}(\nu) g_j^\nu . \tag{6.23}$$

Moreover, the formula

$$y_i^n(\nu) \le (\lambda+\lambda_1)p_{iB}(n-2,\nu) + y_i^{n-2}(\nu) - \sum_{j=0}^{\infty} p_{ij}(n-2,\nu) g_j^\nu \tag{6.24}$$

can be proved easily by induction. From the recurrence relation (6.24), we obtain

$$y_i^{n+2}(\nu) \le y_i^2(\nu)(\lambda+\lambda_1)\sum_{r=1}^{n} p_{iB}(r,\nu) - \sum_{r=1}^{n}\sum_{j=0}^{\infty} p_{ij}(r,\nu) g_j^\nu ,$$

and

$$\begin{aligned}\frac{\sum_{r=1}^{n}\sum_{j=0}^{\infty} p_{ij}(r,\nu) g_j^\nu}{n} &\le \frac{y_i^2}{n} + (\lambda+\lambda_1)\frac{\sum_{r=1}^{n} p_{iB}^r}{n} - \frac{y_i^{n+2}(\nu)}{n} \\ &< \frac{y_i^2(\nu)}{n} + \lambda + \lambda_1 < y_i^2(\nu) + \lambda + \lambda_1 .\end{aligned} \tag{6.25}$$

Take $i \in B$. Then

$$\sum_{j=0}^{\infty}\left(\frac{\sum_{r=1}^{n} p_{ij}(r,\nu)}{n} g_i^\nu\right) < y_i^2(\nu) + \lambda + \lambda_1 < 2\lambda + \lambda_1 . \tag{6.26}$$

Take an arbitrary $M > 0$. Consider the set

$$B^M = \{i : \min_{\nu\in D} g^\nu(i) < M\} .$$

Since $g_i^\nu \to \infty$ uniformly in $\nu \in D$ as $i \to \infty$, the set B^M is finite. Furthermore, if $j \in \mathcal{A}\backslash B^M$, then for any $\nu \in D$ we have $g_j^\nu \ge M$. It follows from (6.26) that

$$\sum_{j\in\mathcal{A}\backslash B^M} \frac{\sum_{r=1}^{n} p_{ij}(r,\nu)}{n} g_j^\nu < 2\lambda + \lambda_1 .$$

Thus we get

$$\sum_{j\in\mathcal{A}\backslash B^M} \frac{\sum_{r=1}^{n} p_{ij}(r,\nu)}{n} < \frac{2\lambda+\lambda_1}{M} . \tag{6.27}$$

Now, it follows from (6.27) that

$$\sum_{j\in B^M} \frac{\sum_{r=1}^{n} p_{ij}(r,\nu)}{n} > 1 - \frac{2\lambda+\lambda_1}{M} . \tag{6.28}$$

We have

$$\pi_j(\nu) = \lim_{n\to\infty} \frac{\sum_{r=1}^{n} p_{ij}^r(\nu)}{n} .$$

Therefore, from the finiteness of B^M and (6.28) it follows that

$$\sum_{j\in \mathcal{A}\setminus B^M} \pi_j(\nu) < \frac{2\lambda+\lambda_1}{M} . \tag{6.29}$$

Since the number M can be chosen arbitrarily large and we can construct the set B^M for it, (6.29) implies the compactness of the family $\{\pi_j(\nu)\}$ of distributions for $\nu \in D$, which implies the continuity of $\pi_j(\nu)$ for any $j \in \mathcal{A}$ and $\nu \in D$ by theorem 6.1.5. The theorem is proved. ■

Theorem 6.2.2 *Assume that the following conditions are satisfied for some $\delta > 0$, some $\gamma > 1$, and a finite nonempty set $B \subset \mathcal{A}$:*

(i) $\sum_{j=0}^{\infty} p_{ij}(\nu) f_j^\nu - f_i^\nu < -\delta, \; i \notin B, \; \nu \in D.$

(ii) $\sup_{i\in B,\nu\in D} \sum_{j=0}^{\infty} p_{ij}(\nu)(f_j^\nu)^\gamma = \lambda_\gamma < \infty.$

(iii) $\sup_{i\in B,\nu\in D} \sum_{j=0}^{\infty} p_{ij}(\nu)|f_j^\nu - f_i^\nu|^\gamma = C_\gamma < \infty.$

(iv) *$f_i^\nu \to \infty$ uniformly in $\nu \in D$ as $i \to \infty$.*

Then the chains L^ν are ergodic for every $\nu \in D$, and the stationary probabilities $\pi_j(\nu)$ are continuous in ν for $\nu \in D$ and $j \in \mathcal{A}$.

Proof For some $\gamma > 1$ let condition (iii) of theorem 6.2.2 be satisfied. Then for any γ_0 such that $1 \le \gamma_0 < \gamma$ we have

$$\sup_{i\in \mathcal{A},\nu\in D} \sum_{j=0}^{\infty} p_{ij}(\nu)|f_j^\nu - f_i^\nu|^{\gamma_0} < C_\gamma + 1 < \infty .$$

Therefore, without loss of generality we may assume that $1 < \gamma \le 2$. For any such γ and any $y, x > 0$ we prove the auxiliary inequality

$$y^\gamma - x^\gamma \le |y-x|^\gamma + x^{\gamma-1}\gamma(y-x) . \tag{6.30}$$

Set $z = y/x$. The inequality (6.30) can be rewritten as

$$z^\gamma - 1 - |z-1|^\gamma - \gamma(z-1) \le 0\,. \tag{6.31}$$

For the proof of (6.31) we consider two cases.

(i) Let $z \ge 1$. Then for $z = 1$ the left side of (6.31) is equal to zero, and for $z > 1$ we have

$$\frac{d(z^\gamma - 1 - (z-1)^\gamma - \gamma(z-1))}{dz} = \gamma[z^{\gamma-1} - (z-1)^{\gamma-1} - 1] < 0\,.$$

Consequently, (6.31) is satisfied for $z \ge 1$.

(ii) Let $z \ge 1$. Then inequality (6.31) turns into an equality for $z = 1$, and for $z < 1$ we have

$$\frac{d(z^\gamma - 1 - (1-z)^\gamma - \gamma(z-1))}{dz} = \gamma[z^{\gamma-1} + (1-z)^{\gamma-1} - 1] > 0\,.$$

Consequently, (6.31) is satisfied for $z < 1$, and so is inequality (6.30), for any $1 \le \gamma \le 2$ and $y, x > 0$.

Let us use (6.30) to estimate $\sum_{j=0}^{\infty} p_{ij}(\nu)[(f_j^\nu)^\lambda - (f_i^\nu)^\gamma]$, for $i \notin B$. We have

$$\sum_{j=0}^{\infty} p_{ij}(\nu)[(f_j^\nu)^\gamma - (f_i^\nu)^\gamma]$$
$$\le \sum_{j=0}^{\infty} p_{ij}(\nu)[|f_j^\nu - f_i^\nu|^\gamma + (f_i^\nu)^{\gamma-1}\gamma(f_j^\nu - f_i^\nu)]$$
$$\le \sum_{j=0}^{\infty} p_{ij}(\nu)|f_j^\nu - f_i^\nu|^\gamma + \gamma(f_i^\nu)^{\gamma-1}\sum_{j=0}^{\infty} p_{ij}(\nu)(f_j^\nu f_i^\nu).$$

Taking into account conditions (i) and (iii) in theorem 6.2.2, we finally obtain the estimate

$$\sum_{j=0}^{\infty} p_{ij}(\nu)[(f_j^\nu)^\gamma - (f_i^\nu)^\gamma] \le C_\gamma - \gamma(f_i^\nu)^{\gamma-1}\delta\,. \tag{6.32}$$

If for some family $\{f_i^\nu\}$ of functions the hypotheses of theorem 6.2.2 are satisfied, then they will also be satisfied for the family $\{f_i^\nu + r\} = \{\tilde{f}_i^\nu\}$, where $r > 0$ is arbitrary. (The second hypothesis of the theorem will be satisfied with another constant $\tilde{\lambda}_\nu < \infty$.) Therefore, without loss of generality we may assume that $\gamma(f_i^\nu)^{\gamma-1}\delta - c_\gamma > \sigma > 0$ for some σ and any $i \in \mathcal{A}$, $\nu \in D$. Set $\tilde{f}_i^\nu = (f_i^\nu)^\gamma$ and $\tilde{g}_i^\nu = \gamma(f_i^\nu)^{\gamma-1}\delta - c_\gamma$. Using (6.32) and condition (ii) in the theorem, we obtain

$$\sum_{j=0}^{\infty} p_{ij}(\nu)\tilde{f}_j^\nu - \tilde{f}_i^\nu \quad < \quad -\tilde{g}_i^\nu, \quad i \notin B,\ \nu \in D\,,$$

$$\sum_{j=0}^{\infty} p_{ij}(\nu)\tilde{f}_j^\nu \quad < \quad \tilde{\lambda}_\gamma < \infty, \; i \in B, \; \nu \in D .$$

Since $\gamma > 1$, the fact that the f_i^ν tend to infinity uniformly as $i \to \infty$ implies the same for $g_i^\nu = \gamma(f_i^\nu)^{\gamma-1}\delta - c_\gamma$. Hence, for the functions $\{\tilde{f}_i^\nu\}$ and $\{\tilde{g}_i^\nu\}$ all hypotheses of theorem 6.2.1 are satisfied. Consequently, the $\pi_j(\nu)$ are continuous in ν, for $\nu \in D$ and $j \in \mathcal{A}$. The theorem is proved. ■

Remark Let $\xi_0, \xi_1, \ldots$ be the sequence of random variables corresponding to the Markov chain L. Let C be a set, $C \subset \mathcal{A} = \{0, 1, \ldots\}$. For $i \in C$ introduce

$$f_{iC}^n = P(\xi_i \notin c, \ldots, \xi_{n-1} \notin C, \; \xi_n \in C / \xi_0 = i) .$$

Let τ be the stopping-time representing the epoch of the first entry to the set C. Then, under the condition $\xi_0 = i$, we have, for any $\gamma > 0$,

$$E(\tau^\gamma) = \sum_{n=1}^{\infty} n^\gamma f_{iC}^n .$$

When $E(\tau^\gamma)$ is finite, we will speak of the *γ-recurrence* of the set C. As follows from results of [Kal73], the hypotheses of theorem 6.2.2 guarantee the γ-recurrence of B, uniformly in $\nu \in D$. Moreover, it can be shown that the uniform γ-recurrence of the family of chains $\{L^\nu\}$ with $\gamma > 1$ implies the continuity of the stationary probabilities. In this way, we have outlined yet another method of proving theorem 6.2.2.

On the set $\mathcal{A} = \{0, 1, \ldots\}$ let a family of integral-valued, positive, uniformly bounded functions $k^\nu = \{k_i^\nu\}$ be given for which

$$\sup_{i \in \mathcal{A}, \nu \in D} k_i^\nu = b < \infty .$$

Concerning the function $\{f_i^\nu\}$ already introduced, we assume the following condition to be satisfied.

Boundedness condition There is a $d > 0$ such that

$$\sup_{\nu \in D} |f_i^\nu - f_j^\nu| > d \quad \text{implies that } p_{ij}(\nu) = 0 .$$

Theorem 6.2.3 *Assume that the inequalities*

$$\sum_{j=0}^{\infty} p_{ij}(k_i^\nu, \nu) f_j^\nu - f_i^\nu < -\epsilon \tag{6.33}$$

are satisfied for some $\epsilon > 0$, all $\nu \in D$, and all i except some finite nonempty set B. Then the chains L^ν are ergodic for every $\nu \in D$, and the stationary probabilities $\pi_j(\nu)$ are continuous in ν for $\nu \in D$ and $j \in \mathcal{A}$.

We first proceed to derive two lemmas. Let $\{\xi_i^\nu\}$ be the sequence of random variables corresponding to L^ν. We have

$$f_{ij}^n(\nu) = P(\xi_1^\nu \neq j, \ldots, \xi_{n-1}^\nu \neq j,\ \xi_n^\nu = j | \xi_0^\nu = i\} .$$

Lemma 6.2.4 *The functions $f_{ij}^n(\nu)$ are continuous in $\nu\,(\nu \in D)$, for any natural number n and any $i, j \in \mathcal{A}$.*

Proof We prove the lemma by induction. For $n = 1$ the function $f_{ij}^1(\nu) = p_{ij}(\nu)$ is continuous in ν for any i and j. Assume that $f_{ij}^n(\nu)$ is continuous in ν for any $i, j \in \mathcal{A}$. Use the formula

$$f_{ij}^{n+1}(\nu) = \sum_{k=0}^{\infty} p_{ik}(\nu) f_{kj}^n(\nu) - p_{ij}(\nu) f_{ji}^n(\nu) .$$

As in lemma 6.1.1 of the preceding section, we can prove the uniform convergence of the series $\sum_{k=0}^{\infty} p_{ik}(\nu) f_{kj}^n(\nu)$, which consists of continuous functions. This implies the continuity of $f_{ij}^{n+1}(\nu)$. ■

Lemma 6.2.5 *Let the hypotheses of theorem 6.2.3 be satisfied. Then, for any points $i_0 \in \mathcal{A}$ and $\nu_0 \in D$ chosen beforehand, there exist functions $\{\tilde{k}_i^\nu\}$ and $\{\tilde{f}_i^\nu\}$ such that $\sup_{i \in \mathcal{A}, \nu \in D} \tilde{k}_i^\nu = \tilde{b} < \infty$, and the boundedness condition is satisfied for the functions $\{f_i^\nu\}$ with some constant $\tilde{d} > 0$ and, in place of (6.33), the inequality*

$$\sum_{j=0}^{\infty} p_{ij}(\tilde{k}_i^\nu, \nu) \tilde{f}_i^\nu < -\epsilon_1$$

holds for some $\epsilon_1 > 0$, all $\nu \in D$, and all $i \in \mathcal{A}$ except $i = i_0$.

Proof Without loss of generality we may set $i_0 = 0$. From the assumption that L^{ν_0} has a single essential class of states, it follows for any point $j \in B$ that there exist a positive integer $r_j(\nu_0)$ and $\epsilon_1 > 0$ such that $p_{j0}(r_j(\nu_0), \nu_0) > \epsilon_1$. Since $p_{ij}(n, \nu)$ is a continuous function of ν ($\nu \in D$, and i, j and n are arbitrary), there exists a neighbourhood D_j

of ν_0 such that $p_{j0}(r_j(\nu_0),\nu) > \epsilon_2 = \epsilon_1/2$, for any $\nu \in \tilde{D}_j$. Let

$$\begin{aligned}
\tilde{D} &= \bigcap_{j\in B} \tilde{D}_j \,, \\
\tilde{k}_j^\nu &= \begin{cases} k_j^\nu, & j \notin B,\ \nu \in \tilde{D}\,, \\ r_j^{\nu_0}, & j \in B,\ \nu \in \tilde{D}\,, \end{cases} \\
\tilde{f}_j^\nu &= \begin{cases} f_j^\nu, & j \neq 0,\ \nu \in \tilde{D}\,, \\ f_0^\nu - 2\,\dfrac{d\tilde{b}}{\epsilon_2}, & j = 0,\ \nu \in \tilde{D}\,, \end{cases}
\end{aligned}$$

where $\tilde{b}$ is defined in the following way:

$$\sup_{i\in\mathcal{A},\nu\in D} \tilde{k}_j^\nu = \tilde{b} < \infty, \quad \tilde{d} = d + 2\frac{db}{\epsilon_2}\,.$$

For any $i \notin B \cup 0$ we have

$$\sum_{j\in\mathcal{A}} p_{ij}(\tilde{k}_i^\nu,\nu)\tilde{f}_j^\nu - \tilde{f}_i^\nu < -\epsilon \ \ (\nu \in \tilde{D})\,. \tag{6.34}$$

For $i \in B\backslash 0$ and $\nu \in \tilde{D}$ we have

$$\sum_{j=0}^{\infty} p_{ij}(\tilde{k}_i^\nu,\nu)\tilde{f}_j^\nu - \tilde{f}_i^\nu = \sum_{j=0}^{\infty} p_{ij}(\tilde{k}_i^\nu,\nu)f_j^\nu - p_{i0}(\tilde{k}_i^\nu,\nu)\frac{2d\tilde{b}}{\epsilon_2} - f_i^\nu \tag{6.35}$$

$$< f_i^\nu + d\tilde{k}_i^\nu - 2d\tilde{b} - f_i^\nu < -d\tilde{b}\,. \tag{6.36}$$

Hence, if we set $\tilde{\epsilon} = \min\,(\epsilon, d\tilde{b})$, we finally obtain

$$\sum_{j=0}^{\infty} p_{ij}(\tilde{k}_i^\nu,\nu)\tilde{f}_j^\nu - \tilde{f}_i^\nu < -\tilde{\epsilon} \quad (\nu \in \tilde{D},\ i \neq 0)\,. \tag{6.37}$$

The proof of the lemma is completed. ■

Proof of theorem 6.2.3 The ergodicity of the chains L^ν $\nu \in D$ follows from theorem 2.2.3. Let us fix any point $i_0 \in \mathcal{A}$. Without loss of generality, let $i_0 = 0$. We prove that $\pi_0(\nu)$ is a continuous function of ν for $\nu \in D$. Take an arbitrary point $\nu_0 \in D$. It follows from lemma 6.2.5 that we can correct the functions f_i^ν and k_i^ν so that for some neighbourhood $\tilde{D}$ of ν_0 we have (6.37). We may assume that $\tilde{f}_0^\nu < f_i^\nu$, for any $i \in \mathcal{A}$ and $\nu \in \tilde{D}$, since otherwise this can be achieved easily by decreasing the values of $\tilde{f}_0^\nu$; in this case the inequalities (6.37) are not violated. We introduce the quantity

$$m_0(\nu) = \sum_{n=1}^{\infty} n f_{00}^n(\nu)\,,$$

which is the mean time of entrance of L^ν into the null state. It follows from the ergodicity of the chains $L^\nu(\nu \in D)$ that $m_0(\nu)$ is finite, and

$$\pi_0(\nu) = \frac{1}{m_0(\nu)} .$$

We show that $m_0(\nu)$ is continuous in ν for $\nu \in D$. From the sequence of random variables $\xi_0^\nu = 0, \xi_1^\nu, \ldots$ corresponding to L^ν we form a random sequence $\{S_n^\nu\}$ by setting $S_n^\nu = f^\nu\{\xi_n^\nu\}$. From the sequence $\{\xi_i^\nu\}$ we also form an integral-valued sequence $\{N_i^\nu\}$ by setting $N_0^\nu = \tilde{k}^\nu(\xi_0^\nu)$ and $N_i^\nu = N_{i-1}^\nu + \tilde{k}^\nu(\xi_{i-1}^\nu)$. It follows from the uniform boundedness of the functions $\tilde{k}_i^\nu(i \in \mathcal{A},\ \nu \in D)$ that $1 \le N_{i+1}^\nu - N_i^\nu \le \tilde{b}$, for any $i \in \mathcal{A}$ and $\nu \in D$, with probability 1. It follows from (6.37) that

$$E(S_{N_i}^\nu / S_{N_{i-1}}^\nu > \tilde{f}_0^\nu) < S_{N_{i-1}}^\nu - \epsilon \quad \text{a.s.} \tag{6.38}$$

We have

$$\begin{aligned} f_{00}^n(\nu) &= P(\xi_1^\nu \neq 0, \ldots, \xi_{n-1}^\nu \neq 0,\ \xi_n^\nu = 0 | \xi_0^\nu = 0) \\ &= P(S_1^\nu \ge \tilde{f}_0^\nu, \ldots, S_{n-1}^\nu > \tilde{f}_0^\nu,\ S_n^\nu = \tilde{f}_0^\nu | S_0^\nu = \tilde{f}_0^\nu) . \end{aligned}$$

Therefore, taking into account (6.38) and applying theorem 2.1.7, we obtain the following estimate for $f_{00}^n(\nu)$:

$$f_{00}^n(\nu) < c\exp(-\delta n), \quad n \in \mathcal{A}, \nu \in \tilde{D} , \tag{6.39}$$

where $c, \delta > 0$ are constants not depending on ν. Taking account of (6.39), we conclude that the series $\sum_{n=1}^\infty n f_{00}^n(\nu)$, which consists of continuous functions (see lemma 6.2.4), converges uniformly in ν for $\nu \in \tilde{D}$. Therefore the sum $m_0(\nu)$ of the series is a continuous functions of ν, for $\nu \in D$, which in turn implies the continuity of $\pi_0(\nu)$ in ν for $\nu \in \tilde{D}$. It remains to note that $i = 0$ and $\nu_0 \in D$ were chosen arbitrarily. Hence the $\pi_j(\nu)$ are continuous functions of the parameter ν, for $\nu \in D$ and all $j \in \mathcal{A}$. The proof of the theorem is finished. ∎

6.3 Continuity of random walks in $\mathbf{Z}_+^N$

Consider a family $\{L^\nu\}$ of homogeneous irreducible aperiodic Markov chains in discrete time, with state space $\mathbf{Z}_+^N$. Here $\nu \in D$, where D is an open subset of the real line. $p_{\alpha\beta}(t, \nu)$, for $\alpha, \beta \in \mathbf{Z}_+^N$, is the probability of the transition of L^ν from the point α to the point β in t steps. Concerning the family $\{L^\nu\}$ of random walks, we assume that the homogeneity condition and the boundedness of jumps hold uniformly in $\nu \in D$. Let $B_{ct}^\wedge$ be the sets introduced in section 4.3. We assume the following.

Homogeneity condition: There is a $c > 0$ such that, for any Λ and any vector $a = (a_1, \ldots, a_N)$, with $a_i \geq 0, 1 \leq i \leq N$, and $a_j = 0$, for $j \notin \Lambda$, we have

$$p_{\alpha\beta}(\nu) = p_{\alpha+a,\beta+a}(\nu) ,$$

for all $\alpha \in B_{cc}^{\Lambda} \cap \mathbf{Z}_+^N$ and $\beta \in \mathbf{Z}_+^N$ and $\nu \in D$.

Boundedness of the jumps: For any α, the number of β 's such that $\sup_{\nu \in p_{\alpha\beta}}(\nu) > 0$ is finite

The boundedness conditioin is equivalent to the following: There exists $d > 0$ such that $\| \alpha - \beta \| > d$ implies $p_{\alpha\beta}(\nu) = 0$ for any ν. As before, we shall assume that the $p_{\alpha\beta}(1, \nu)$ are continuous functions of ν for any $\alpha, \beta \in \mathbf{Z}_+^N$ and $\nu \in D$. For every chain L^ν let us construct a vector field V^ν by the method indicated in chapter 4. Then we obtain a family $\{V^\nu\}$ of vector fields.

We shall say that the *family* $\{V^\nu, (\nu \in \tilde{D} \subset D)\}$ satisfies *Condition* $\tilde{B}$ if, for some δ, b, and $p > 0$, there exists a function $f(\alpha)$, $\alpha \in \mathbf{R}_+^N$ satisfying the following conditions:

(i) $f(\alpha) > 0, \quad \alpha \in R_+^N$.

(ii) $f(\alpha) - f(\beta) \leq b \| \alpha - \beta \|$ for any $\alpha, \beta \in R_+^N$.

(iii) For any Λ either all the $L^\Lambda(\nu)$ $(\nu \in \tilde{D})$ or none of them are ergodic.

(iv) For any Λ such that $L^\Lambda(\nu)$ is ergodic and for $\Lambda = \{1, \ldots, N\}$ and all $\alpha \in B^\Lambda \cap B_{pp}^\Lambda$ we have

$$\sup_{\nu \in \tilde{D}} (f(\alpha + v^\nu(\alpha)) - f(\alpha)) < -\delta .$$

Theorem 6.3.1 *If there exists a set $U \subset D$ such that $\{V^\nu, \nu \in U\}$, satisfies Condition $\tilde{B}$, then, for all $\nu \in U$, the chains L^ν are ergodic, and the stationary probabilities $\pi_\alpha(\nu)$ are continuous in ν for any $\alpha \in \mathbf{Z}_+^N$ and $\nu \in U$.*

Proof Assume that there exist a set $U \subset D$ and a function $f(\alpha), \alpha \in R_+^N$, such that Condition $\tilde{B}$ is satisfied. Set $f^\nu \equiv f$, i.e. set $f^\nu(\alpha) = f(\alpha)$ for every $\alpha \in \mathbf{Z}_+^N$ and $\nu \in U$. It follows from the proof of theorem 6.1.3 that for any $\nu \in U$ there is a function $m^\nu(\alpha)$ such that

$$\sup_{\alpha \in Z_+^N} m^\nu(\alpha) = m^\nu < \infty$$

and, for all $\alpha \in \mathbf{Z}_+^N$ except some finite set C^ν,

$$\sum_{\beta \in Z_+^N} p_{\alpha\beta}(m(\alpha), \nu) f_\beta^\nu - f_\alpha^\nu < -\epsilon_1(\nu) \tag{6.40}$$

for some $\epsilon_1(\nu) > 0$. From the method of proof of theorem 4.3.4, it follows that

$$\begin{cases} \sup_{\nu \in U} m^\nu < \infty \,, \\ \inf_{\nu \in U} \epsilon_1(\nu) > 0 \,, \end{cases} \tag{6.41}$$

and $\bigcup_{\nu \in U} C^\nu$ is a finite set.

From (6.40) and (6.41) we conclude that all the hypotheses of theorem 6.2.3 are satisfied. Consequently, all chains L^ν are ergodic for $\nu \in U$, and the stationary probabilities $\pi_\alpha(\nu)$ are continuous in ν for $\nu \in U$ and any $\alpha \in \mathbf{Z}_+^N$. The theorem is proved. ■

For a random walk $\mathcal{L}$ in $\mathbf{Z}_+^N$, where $N \leq 3$, in chapters 3 and 4 a method was given for constructing the function $f(\alpha)$ satisfying Condition **B**, which leads to the formulation of ergodicity conditions in terms of random walks of lower dimensions. Now we prove a theorem showing that the satisfaction of these ergodicity conditions for the chain L^{ν_0} guarantees the continuity of the stationary probabilities of $\{L^\nu\}$ in some neighbourhood of ν_0. We formulate the theorem for random walks in $\mathbf{Z}_+^3$. (This can be done analogously for $\mathbf{Z}_+^1$ or $\mathbf{Z}_+^2$.)

Theorem 6.3.2 *Assume that the Markov chain L^{ν_0}, where $\nu_0 \in D$, satisfies the hypotheses of theorem 4.4.4 which guarantee the ergodicity of L^{ν_0}. Then there exists a neighbourhood $U \subset D$ of the point ν_0 such that for all $\nu \in U$ the chains L^ν are ergodic, and the stationary probabilities $\pi_\alpha(\nu)$ are continuous in ν for any $\alpha \in \mathbf{Z}_+^3$ and $\nu \in U$.*

Proof It follows from the proof of theorem 4.4.4 that there exists a function $f(\alpha)$, $\alpha \in \mathbf{R}_+^3$, which satisfies Condition **B** introduced in section 4.3 of chapter 4. In particular,

(i) $f(\alpha) > 0, \quad \alpha \in \mathbf{R}_+^3$.
(ii) $|f(\alpha) - f(\beta)| \leq b \parallel \alpha - \beta \parallel$, $\alpha, \beta \in \mathbf{R}_+^3$, where $b > 0$ is a constant.

Condition **B** guarantees the existence of a function $m(\alpha)$, $\alpha \in \mathbf{Z}_+^3$, with values in the set of natural numbers such that

$$\sup_{\alpha \in Z_+^3} m(\alpha) = m < \infty$$

and, for all $\alpha \in \mathbf{Z}_+^3$ except some finite set R,

$$\sum_{\beta \in Z_+^3} p_{\alpha\beta}(m(\alpha), \nu_0) f_\beta - f_\alpha < -\epsilon \tag{6.42}$$

for some $\epsilon > 0$. From the homogeneity condition and the boundedness of jumps for the family of random walks $\{L^\nu\}$ it follows that $p_{\alpha\beta}(t, \nu)$ is a continuous function of ν for $\nu \in D$ uniformly in $\alpha, \beta \in \mathbf{Z}_+^3$ (t is an arbitrarily given natural number). Therefore, taking account of properties (i) and (ii) of the function $f(\alpha)$ we conclude that there is a neighbourhood U of ν_0 such that for all $\nu \in U$ and $\alpha \in \mathbf{Z}_+^3 \backslash B$

$$\sum_{\beta \in Z_+^3} p_{\alpha\beta}(m(\alpha), \nu) f_\beta - f_\alpha < \frac{\epsilon}{2} \, . \tag{6.43}$$

From the properties of $f(\alpha)$ and the uniform boundedness of the jumps of the random walks in $\{L^\nu\}$, it follows that there is a $d > 0$ such that

$$\sup_{\nu \in D} |f_j^\nu - f_i^\nu| > d \quad \text{implies that } p_{ij}(\nu) = 0 \, .$$

Hence all the hypotheses of theorem 6.2.3 are satisfied. Consequently, the chains L^ν are ergodic for all $\nu \in U$, and the stationary probabilities $\pi_\alpha(\nu)$ are continuous in ν, for $\nu \in U$ and any $\alpha \in \mathbf{Z}_+^3$. The theorem is proved. ∎

7

Exponential convergence and analyticity for ergodic Markov chains

The aim of this chapter is to prove theorems of the following type: if the Lyapounov function in Foster's criterion for ergodicity satisfies certain conditions, then

(i) the convergence to the stationary distribution is exponentially fast; more precisely,

$$| p_{ij}^{(n)} - \pi_j | < C_i e^{-\delta n} \, , \tag{7.1}$$

where C_i depends on i but $\delta > 0$ does not depend on i or on j;

(ii) π_j decrease exponentially with j, for some metric defined on the state space, i.e.

$$\pi_j < C e^{-f(j)}, \tag{7.2}$$

for some function $f(j)$;

(iii) the ergodicity is preserved under small perturbations of the transition probabilities and the stationary probabilities depend analytically on these perturbations.

7.1 Analytic Lyapounov families

Let us consider a family of Markov chains $\{L^\nu, \nu \in \mathcal{D}\}$ — where $\mathcal{D}$ represents an interval containing 0 — defined on the same state space S. The matrix $P_\nu = [p_{ij}(\nu)]_{i,j \in S}$ of transition probabilities can be considered as a bounded linear operator in $l_1(S)$. Let us assume that P_ν is analytic in ν as a function in $\mathcal{D}$ taking its value in the Banach algebra of all bounded operators in $l_1(S)$. This means that P_ν can be Taylor expanded as

$$P_\nu = \sum_{n=0}^{\infty} P_n \nu^n \ , \tag{7.3}$$

where the P_n's are bounded linear operators satisfying

$$\|P_n\| \le C a^n, \tag{7.4}$$

for some $C, a > 0$, i.e. the series in (7.3) is norm-convergent.

Definition 7.1.1 *We say that in this case we have an* analytic family *of Markov chains. We say that this family is an* analytic Lyapounov family *if in addition the following conditions hold: there exist nonnegative functions f_i^ν, and positive integer-valued functions $k_i^\nu, \nu \in \mathcal{D}$, $i \in S$, such that*

(i) $\sup_{i \in S, \nu \in \mathcal{D}} k_i^\nu = b < \infty$;
(ii) there exist $C, \mu > 0$ such that $f_i^\nu > C i^\mu$, for any i and $\nu \in \mathcal{D}$;
(iii) there exists $d > 0$ such that

$$p_{ij}(l, \nu) = 0, \ \textit{for all } \nu \in \mathcal{D}, \ \textit{whenever } \mid f_i^\nu - f_j^\nu \mid > d \ ,$$

where the $p_{ij}(l,\nu)$'s denote the one-step transition probabilities corresponding to the parameter ν;
(iv) there exist $n > 0$ and $\delta > 0$ such that, for any $i \in S$ and any $j \in V_i \stackrel{\text{def}}{=} \{j : \sup_{\nu \in \mathcal{D}} p_{ji}\nu) > 0\}$,

$$p_{ji}(n, 0) > \delta; \tag{7.5}$$

(v) for all $\nu \in \mathcal{D}, i \in S$ except some finite $B \subset S$ and some $\epsilon > 0$,

$$\sum_{j \in S} p_{ij}(k_i^\nu, \nu) - f_i^\nu < -\epsilon, \tag{7.6}$$

so that, by Foster's criterion the L_ν's are ergodic for any ν.

We will also say that an MC is an analytic Lyapounov MC if the family $L_\nu \equiv L_0$ is analytic Lyapounov.

Theorem 7.1.2 *If L_ν is an analytic Lyapounov family then there exists $\nu_0 > 0$ such that*

(i) *there exist $C_2, \delta_2 > 0$ such that*

$$\pi_i(\nu) < C_2 \exp(-\delta_2 f_i^\nu), \tag{7.7}$$

for all $i \in S$, $\nu \in \mathcal{D}$;

(ii) *there exist constants* $\sigma_2, C_6, \delta_6 > 0$ *such that*

$$\sum_{j \in S} | p_{ij}(n,\nu) - \pi_j(\nu) | < C_6 \exp(-\delta_6 n), \text{for all } \nu \in \mathcal{D}, i \in S,\ n > \sigma_2 f_i^\nu \ ; \tag{7.8}$$

(i) *the stationary probabilities* $\pi_i(\nu)$ *are analytic in* ν*, for* $| \nu | < \nu_0$ *and all* $i \in S$.

Corollary 7.1.3 *If there is no dependence on* ν*, i.e.*

$$L_\nu \equiv L_0,\ p_{ij}(\nu) \equiv p_{ij}(0) \equiv p_{ij},\ \pi_i(\nu) \equiv \pi_i,$$

and if all properties (i)–(v) in definition 7.1.1 hold for $\nu = 0$*, then assertions* (i) *and* (ii) *hold, i.e., for all* $n > \sigma_2 f_i^\nu$,

$$\pi_i \le C_2\, l^{-\delta_2 n} \tag{7.9}$$

and there exist constants $C, \delta_1, \delta_2 > 0$ *such that, for any* i*, we have*

$$\sum_{j \in S} | p_{ij}^{(n)} - \pi_j | < C\, l^{\delta_1 f_i - \delta_2 n}\ . \tag{7.10}$$

We shall now make some comments about the nature of this exponential convergence. It is well known that uniform exponential convergence, i.e.

$$\sum_j | \pi_j - p_{ij}^{(n)} | < C\, l^{-\delta n}\ , \tag{7.11}$$

for some $C, \delta > 0$ not depending on i, holds for any finite MC and more generally for MCs satisfying Doeblin's condition defined in chapter 1. The difference between (7.11) and (7.10) is connected with the nature of the spectrum of the operator P in $l_1(S)$. Case (7.11) suggests that the eigenvalue 1 is isolated, while in the case (7.10) this eigenvalue is generally embedded in an (absolutely) continuous spectrum. The simplest example of this latter situation is given by the maximally homogeneous random walk in $\mathbf{Z}_+^1$. The complete structure of the spectrum of P can be easily obtained by means of generating functions.

7.2 Proof of the exponential convergence

First we shall prove the assertion (i) of theorem 7.1.2.

The ergodicity of the chains L^ν and the continuity of the stationary probabilities $\pi_j(\nu)$ $(j \in A;\ \nu \in D)$ follow from theorem 6.2.3. Let

$\xi_0^\nu, \xi_1^\nu, \ldots$ be the sequence of random variables corresponding to L^ν. We introduce the classical *taboo* quantities

$$\begin{aligned} f_{ij}^n(\nu) &= P(\xi_1^\nu \neq j,\ \xi_2^\nu \neq j, \ldots,\ \xi_{n-1}^\nu \neq j,\ \xi_n^\nu = j/\xi_0^\nu = i)\ , \\ {}_k p_{ij}^n(\nu) &= P(\xi_1^\nu \neq k,\ \xi_2^\nu \neq k, \ldots,\ \xi_{n-1}^\nu \neq k,\ \xi_n^\nu = j/\xi_0^\nu = i)\ , \\ {}_k p_{ij}^*(\nu) &= \sum_{n=1}^{\infty} {}_k p_{ij}^n(\nu)\ . \end{aligned}$$

In theorem 6.2.3, the estimate

$$f_{00}^n(\nu) < c_1 \exp(-\delta_1 n) \tag{7.12}$$

has been proved for some $c_1, \delta_1 > 0$ and any $n \in A$ and $\nu \in D$. We can also prove that

$${}_0 p_{0j} < c_1 \exp(-\delta_1 n)\ , \tag{7.13}$$

in an entirely analogous way. For the irreducible aperiodic recurrent chain L^ν, we have (see [Chu67])

$$\lim_{m\to\infty} \frac{\sum_{n=0}^m p_{ij}(n,\nu)}{\sum_{n=0}^m p_{ii}(n,\nu)} = {}_i p_{ij}^*(\nu)\ . \tag{7.14}$$

For an ergodic chain, (7.14) becomes

$$\frac{\pi_j(\nu)}{\pi_i(\nu)} = {}_i p_{ij}^*(\nu)\ . \tag{7.15}$$

Consequently,

$$\pi_j(\nu) < {}_0 p_{0j}^*(\nu)\ , \tag{7.16}$$

$${}_0 p_{0j}^*(\nu) = \sum_{n=1}^{\infty} {}_0 p_{0j}^n(\nu) = \sum_{n=1}^{[f_j^\nu/d]} {}_0 p_{0j}^n(\nu) + \sum_{[f_j^\nu/d]}^{\infty} {}_0 p_{0j}^n(\nu)\ . \tag{7.17}$$

The first sum in the right-hand side of (7.17) is equal to zero. This follows from point (iii) of definition 7.1.1. From (7.13), it follows that there exist constants c_2, $\delta_2 > 0$ such that

$$\sum_{n=[f_j^\nu/d]}^{\infty} {}_0 p_{0j}^n(\nu) < c_2 \exp(-\delta_2 f_j^\nu)\ . \tag{7.18}$$

So assertion (i) is proved.

Lemma 7.2.1 *There exist constants c_3, $\delta_3 > 0$, such that*

$$| p_{00}(n,\nu) - \pi_0(\nu) | < c_3 \exp(-\delta_3 n), \tag{7.19}$$

for any $n \in A$ and $\nu \in D$.

Proof Introduce the generating functions

$$F_{ij}^{\nu}(z) = \sum_{n=0}^{\infty} f_{ij}^{n}(\nu) z^n , \tag{7.20}$$

$$\mathbf{P}_{ij}^{\nu}(z) = \sum_{n=0}^{\infty} p_{ij}(n,\nu) z^n . \tag{7.21}$$

Using the relation (convolution)

$$p_{ij}(n,\nu) = \sum_{s=1}^{n} f_{ii}^{s}(\nu) p_{ii}(n-s,\nu)$$

for the generating functions, we obtain

$$\mathbf{P}_{ii}^{\nu}(z) = \frac{1}{1 - F_{ii}^{\nu}(z)} \quad , \quad \text{valid for } i \geq 0 .$$

It follows from (7.12) that $F_{00}^{\nu}(z)$ is analytic in z, for $| z | < 1 - \sigma$, for some $\sigma > 0$ and any $\nu \in D$. Moreover, $| F_{ii}^{\nu}(z) | < 1$ for $| z | = 1$, $z \neq 1$, since the greatest common divisor k such that $f_{00}^{k}(\nu) \neq 0$ is 1. There exists a neighbourhood U of $z = 1$ such that the equation $F_{00}^{\nu}(z) = 1$ has in U only one root, namely $z = 1$. Therefore, for some $\sigma_1 > 0$, the equation $F_{00}^{\nu}(z) - 1 = 0$ has no other roots for $| z | < 1 + \sigma_1$. Consequently $\mathbf{P}_{00}^{\nu}(z) = 1/[1 - F_{00}^{\nu}(z)]$ is a meromorphic function for $| z | < 1 + \sigma_1$, having exactly one pole of first order at $z = 1$. Let Res(1) be the residue of $\mathbf{P}_{00}^{\nu}(z)$ at $z = 1$. Then

$$\begin{aligned} \text{Res}(1) &= \frac{1}{1 - F_{00}^{\nu}(z)} = \lim_{z \to 1} \frac{z-1}{1 - F_{00}^{\nu}(z)} = \lim_{z \to 1} \frac{1}{(1 - F_{00}^{\nu}(z))/(z-1)} \\ &= -\frac{1}{(dF_{00}^{\nu}(z)/dz)\,|_{z=1}} = -\frac{1}{\sum_{n=1}^{\infty}(1/n f_{00}^{n}(\nu))} = -\pi_0^{\nu} . \end{aligned}$$

Then $\tilde{\mathbf{P}}_{00}^{\nu}(z) = \mathbf{P}_{00}^{\nu}(z) - \pi_0^{\nu}/(1-z)$ is holomorphic for $|z| < 1 + \epsilon_2$. On the other hand, since

$$\tilde{\mathbf{P}}^{\nu}(z) = \sum_{n=0}^{\infty} (p_{00}(n,\nu) - \pi_0) z^n ,$$

there exist constants $c_3, \delta_3 > 0$ such that

$$|p_{00}(n,\nu) - \pi_0(\nu)| < c_3 \exp(-\delta_3 n) .$$

Lemma 7.2.1 is proved. ■

Lemma 7.2.2 *There exist constants $\sigma, c_4, \delta_4 > 0$, such that*

$$|p_{i0}(n,\nu) - \pi_0(\nu)| < c_4 \exp(-\delta_4 n) \tag{7.22}$$

for any $\nu \in D$, $i \in A$, and $n > \sigma f_i^\nu$.

Proof Using theorem 2.1.8, we find, as in the derivation of (7.12), that there exist constants $b_1, a_1, \sigma_1 > 0$, such that

$$f_{i0}^n(\nu) < b_1 \exp(-a_1 n), \tag{7.23}$$

for any $\nu \in D, i \in A$ and $n > \sigma_1 f_i^\nu$. Since

$$p_{i0}(n,\nu) = \sum_{r=1}^{n} f_{i0}^{n-r}(\nu) p_{00}(r,\nu), \tag{7.24}$$

it follows from (7.24) that

$$|p_{i0}(n,\nu) - \pi_0(\nu)| \le \sum_{r=1}^{n} |p_{00}(r,\nu) - \pi_0(\nu)| f_{i0}^{n-r}(\nu) + \pi_0 \sum_{r=n+1}^{\infty} f_{i0}^r(\nu) . \tag{7.25}$$

Choosing $\sigma > \sigma_1$ and $n > \sigma f_i^\nu$, we will at once estimate the right-hand side of (7.24).

(a) Let $r < \epsilon_1 n$, with $(1-\epsilon_1)\sigma > \sigma_1$. Then $(n-r) > (1-\epsilon_1)n > \sigma_1 f_i^\nu$ and, by using (7.23), we obtain

$$f_{i0}^{n-r}(\nu) < b_1 \exp[-a_1(n-r)] < b_1 \exp[-a_1(1-\epsilon_1)n] .$$

(b) Let now $r \ge \epsilon_1 n$. Then lemma 7.2.1 yields

$$|p_{00}(r,\nu) - \pi_0(\nu)| < c_3 \exp(-\delta_3 r) < c_3 \exp(-\delta_3 \epsilon_1 n) .$$

Combining both cases, we obtain

$$\sum_{r=1}^{n} |p_{00}(r,\nu) - \pi_0(\nu)| f_{i0}^{n-r}(\nu) < n b_2 \exp(-a_2 n), \tag{7.26}$$

for any $n > \sigma f_i^\nu$ and some b_2, $a_2 > 0$. It follows from (7.23) that

$$\pi_0 \sum_{r=n+1}^{\infty} f_{i0}^r(\nu) < b_3 \exp(-a_3 n) . \tag{7.27}$$

It follows from (7.26) and (7.27) that there exist constants $c_4, \delta_4 > 0$, such that (7.22) is satisfied for any $\nu \in D$, $i \in A$, and $n > \sigma f_i^\nu$. Lemma 7.2.2 is proved. ■

Lemma 7.2.3 *There exist constants $\sigma_1, c_5, \delta_5 > 0$, such that*

$$|p_{ij}(n,\nu) - \pi_j(\nu)| < c_5 \exp(-\delta_5 n) \tag{7.28}$$

for any $\nu \in D, i$ and $j \in A$ and $n > \sigma_1 f_i^\nu$.

Proof

$$p_{ij}(n,\nu) = \sum_{r=1}^{n} p_{i0}(r,\nu)_0 p_{0j}^{n-r}(\nu) + {}_0p_{ij}^n(\nu)\,, \tag{7.29}$$

$$\begin{aligned} &|\, p_{ij}(n,\nu) - \pi_j(\nu)| \\ &+ \left| \sum_{r=1}^{n} (p_{i0}(r,\nu) - \pi_0(\nu))_0 p_{0i}^{n-r}(\nu) - \pi_0 \sum_{r=n-1}^{\infty} {}_0p_{0j}^r + {}_0p_{ij}^n(\nu) \right| \\ &\leq \sum_{r=1}^{n} |p_{i0}(r,\nu) - \pi_0(\nu)|_0 p_{0j}^{n-r}(\nu) + \pi_0 \sum_{r=n+1}^{\infty} {}_0p_{0j}^r(\nu) + {}_0p_{ij}^n(\nu). \end{aligned} \tag{7.30}$$

Using lemma 7.2.2 and (7.13), we obtain (7.28) directly from (7.30). ■

We are now in a position to prove assertion (ii) of theorem 7.1.2. For that purpose, we write the decomposition

$$\begin{aligned} &\sum_{j=0}^{\infty} |p_{ij}(n,\nu) - \pi_j(\nu)| \\ &= \sum_{j:f_j^\nu > f_i^\nu + nd} |p_{ij}(n,\nu) - \pi_j(\nu)| + \sum_{j:f_j^\nu \leq f_i^\nu + nd} |p_{ij}(n,\nu) - \pi_j(\nu)|. \end{aligned} \tag{7.31}$$

The boundedness of the jumps of the random walks implies that, in the first sum in the right-hand side of (7.31), $p_{ij}(n,0) = 0$, for all j such that $f_j^0 > f_i^0 + nd$. The estimation (7.28) implies therefore that the right side of (7.31) is less than $c_2' \exp\{-\delta_2' f_i^\nu\}$, for some $c_2' > 0$, $\delta_2' > 0$. It follows from lemma 7.2.3 that each term of the second sum is less than $c_5 \exp(-\delta_5 n)$, whenever $n > \sigma_1 f_i^\nu$. Let now M^ν be equal to the number of those j for which

$$f_j^\nu < f_i^\nu + nd\,. \tag{7.32}$$

Let j satisfy (7.32). Then from point (ii) of definition 7.1.1, it follows that

$$cj^\mu < f_j^\nu < f_i^\nu + nd\,,$$

$$j < \left[\frac{1}{c}(f_i^\nu + nd)\right]^{1/\mu} < \left[\frac{1}{c}(\frac{n}{\sigma_1} + nd)\right]^{1/\mu} = n^{1/\mu}\left[\frac{1}{c}(\frac{1}{\sigma_1} + d)\right]^{1/\mu} .$$

Consequently,

$$b_1 M^\nu < n^{1/\nu} , \tag{7.33}$$

where

$$b_1 = \left[\frac{1}{c}(\frac{1}{\sigma_1} + d)\right]^{1/\mu} .$$

Hence

$$\sum_{j: f_j^\nu \leq f_i^\nu + nd} |p_{ij}(n,\nu) - \pi_j(\nu)| < n^{1/\mu} b_1 c_5 \exp(-\delta_5 n) .$$

Therefore, there exist constants $\sigma_2, c_6, \delta_6 > 0$, such that

$$\sum_{j=0}^{\infty} |p_{ij}(n,\nu) - \pi_j(\nu)| < c_6 \exp(-\delta_6 n),$$

whenever $n > \sigma_2 f_i^\nu$. This establishes part (ii) of theorem 7.1.2.

In the next section, we will need the following.

Lemma 7.2.4 *There exist constants $\delta_7, c_7, \sigma_3 > 0$ such that*

$$\sum_{n=1}^{\infty}\sum_{i=0}^{\infty} |p_{ij}(n,\nu) - \pi_j(\nu)| < \sigma_3 f_i^\nu + c_7 \exp(-\delta_7 f_i^\nu) , \tag{7.34}$$

for any $\nu \in D$ and $i \in A$.

Proof We have

$$\sum_{n=1}^{\infty}\sum_{i=0}^{\infty} |p_{ij}(n,\nu) - \pi_i(\nu)| =$$

$$\sum_{n<\sigma_2 f_i^\nu} \sum_{j=0}^{\infty} |p_{ij}(n,\nu) - \pi_j(\nu)| + \sum_{n\geq\sigma_2 f_i^\nu} \sum_{j=0}^{\infty} |p_{ij}(n,\nu) - \pi_j(\nu)| \tag{7.35}$$

together with

$$\sum_{i=0}^{\infty} |p_{ij}(n,\nu) - \pi_j(\nu)| < 2 , \tag{7.36}$$

for any $i, n \in A$. Therefore, the first sum in the right-hand side member of (7.35) is less than $2\sigma_2 f_i^\nu$. To estimate the second sum in the right side of (7.35), we use (7.8), so that

$$\sum_{n\geq\sigma_2 f_i^\nu}\sum_{j=0}^{\infty}|\, p_{ij}(n,\nu)-\pi_j(\nu)| < \sum_{n\geq\sigma_2 f_i^\nu} c_6\exp(-\delta_6 n) < c_7\exp(-\delta_7 f_i^\nu)\,, \tag{7.37}$$

for some $c_7, \delta_7 > 0$. Combining these estimates leads to the assertion of the lemma, which is thus proved. ■

Before the next two sections, which deal with the final part (assertion (i)) of theorem 7.1.2, it is worth making some comments. In particular, it is interesting to note that there exists a *necessary and sufficient* condition for a Markov chain to have the exponential convergence property. First, let us call a Markov chain *geometrically ergodic* if there exist $0 \leq q < 1$ and constants $C_{\alpha\beta} > 0$, for all $\alpha, \beta \in \mathcal{A}$ (the state space),

$$|p_{\alpha\beta} - \pi_\beta| < C_{\alpha\beta} q^n \ , \ n = 1, 2, \ldots \ .$$

Then the following result of Popov (see [Pop77]):

Theorem 7.2.5 *For a Markov chain to be geometrically ergodic, it is necessary and sufficient that there exist a finite set $B \in \mathcal{A}$, $q < 1$ and a non-negative function $f(\alpha)$, $\alpha \in \mathcal{A}$, such that*

$$\begin{aligned}\sum_\beta p_{\alpha\beta} e^{f(\beta)-f(\alpha)} &\leq q\,, \quad \alpha \notin B\,,\\ \sum_\beta p_{\alpha\beta} e^{f(\beta)-f(\alpha)} &\leq \infty\,, \quad \alpha \in B\,.\end{aligned}$$

The proof can be carried out along the same lines as in the lemmas of this section and therefore will not be given in detail. The relationship of this theorem with our results is obvious: we use criteria involving Lyapounov functions with linear growth and here $e^{f(\alpha)}$ increases exponentially fast, in many typical examples. The *sufficient* condition of Popov's result does not bring much with respect to our results and, moreover, it is in general easier to find a function with a linear growth. But the *necessary* condition fills the gap we were feeling: in fact it shows again that one should insist in trying to find a *good* Lyapounov function, whenever one guesses exponential convergence.

7.3 General analyticity theorem

Here we shall state general analyticity conditions for stationary probabilities. At first sight, they might not look very useful, but we show the contrary in the next section, where we use them to prove the last assertion of theorem 7.1.2. Let $\mathcal{X}$ be a Banach space and $B(\mathcal{X})$ the set of all bounded linear operators in $\mathcal{X}$.

Definition 7.3.1 *A set $M \subset B(\mathcal{X})$ is called a set of uniform convergence for the operator $P \in B(\mathcal{X})$ if $PM \subset M$ and there exists a function $\phi(n), n = 1, 2, \ldots$, such that*

$$\phi = \sum_{n=1}^{\infty} \phi(n) < \infty, \tag{7.38}$$

and some element $y \in M$ such that

$$\|P^n x - y\| < \phi(n), \tag{7.39}$$

for all n and $x \in M$.

Theorem 7.3.2 *Let P_ν (sometimes written $P(\nu)$ for notational convenience) depend analytically on ν as a function taking its values in the Banach algebra of operators $B(\mathcal{X})$, and assume that the following conditions are satisfied.*

(i) *For the operator P_0 there exist two sets M_1 and M_2 of uniform convergence such that $M_1 \subset M_2$ and $\inf_{x \in M_2} \|x\| > 0$.*
(ii) *There is a $\nu_0 > 0$ such that $P_\nu x \in M_2$, for all $|\nu| < \nu_0$ and any $x \in M_1$.*
(iii) *There is a $\nu_1 > 0$ such that*

$$x_1 + \frac{(P_\nu - P_0)x_2}{\|P_\nu - P_0\|} \in M_2 \ ,$$

for $|\nu| < \nu_1$ and any $x_1, x_2 \in M_1$.

Then there is a $\nu_2 > 0$ such that, for $|\nu| < \nu_2$ and $x \in M_1$, the limit

$$\lim_{n \to \infty} P^n(\nu)x = r(\nu)$$

exists and depends analytically on ν.

Proof For any $B \in \mathcal{X}(A, \sum)$ and $G \in B(\mathcal{X})$ we write

$$\|G\|_B = \sup_{0 \neq x \in B} \frac{\|G_x\|}{\|x\|} \ .$$

It follows from the hypotheses of the theorem that there exist a function $\phi(n)$ and an element $y \in M_2$ such that conditions (7.38) and (7.39) are satisfied for any $x \in M_2$.

Lemma 7.3.3 *Under the hypotheses of theorem 7.3.2, there is a constant $c > 0$ such that*

$$\|P_0^n(P_\nu - P_0)\|_{M_1} \leq c\phi(n)\|P_\nu - P_0\| \,. \tag{7.40}$$

Proof Take an arbitrary $x \in M_1$. Setting $P_\nu x = z_1$ and $P_0 x = z_2$, we have

$$\begin{aligned}
\| \quad & P_0^n(P_\nu - P_0)x\| \\
= \quad & \|P_0^n(z_1 - z_2)\| \\
= \quad & \left\|P_0^n[z_2\|P_\nu - P_0\| - z_2\|P_\nu - P_0\| + (z_1 - z_2)]\right\| \\
\leq \quad & \|P_\nu - P_0\|\|P_0^n(z_2 - y)\| + \|P_\nu - P_0\|\left\|P_0^n(z_2 + \frac{z_2 - z_1}{\|P_\nu - P_0\|} - y)\right\| \\
\leq \quad & \|P_\nu - P_0\|\phi(n) + \|P_\nu - P_0\|\phi(n) \\
= \quad & 2\phi(n)\|P_\nu - P_0\|.
\end{aligned} \tag{7.41}$$

In the proof of (7.41) we have used the fact that, by assumption, $z_2 = P_0 x$ and $(z_2 - z_1)/\|P_\nu - P_0\|$ belong to the set M_2 of uniform convergence. Moreover, using (7.41), we have

$$\begin{aligned}
\|P_0^n(P_\nu - P_0)\|_{M_1} &= \sup_{x \in M_1} \frac{\|P_0^n(P_\nu - P_0)x\|}{\|x\|} \\
&\leq \frac{2\phi(n)\|P_\nu - P_0\|}{\|x\|} \leq c\phi(n)\|P_\nu - P_0\| \,.
\end{aligned}$$

Lemma 7.3.3 is proved. ■

Let us continue with the proof of theorem 7.3.2. Put

$$Q = Q(\nu) = P_\nu - P_0 \tag{7.42}$$

and write for simplicity

$$Q^{(k,i_1,i_2,\ldots,i_k)} = P_0^{i_1} Q P_0^{i_2} Q \cdots P_0^{i_k} Q \,,$$

where $k \geq 1, i_j \geq 0$ $(j = 1, 2, \ldots, k), \delta$ is the $(k+1)$-tuple $(k, i, \ldots, i_k)$, and $Q^\phi = 1$. Let $r_0 = P_0^\infty x = \lim_{n \to \infty} P_0^n(x)$, for $x \in M_2$. The existence and uniqueness of r_0 obviously follow from conditions (7.38) and (7.39) for the set M_2. We prove that $r_\nu = P_\nu^\infty x = \lim_{n\to\infty} P_\nu^n x$ exists for all $x \in M_1$, is unique and can be represented in the form

$$r_\nu = \sum_\delta Q^\delta r_0 = \sum_{\substack{k\geq 1\\ i_1,\dots,i_k\geq 0}} Q^{(k,i_1,i_2,\dots,i_k)} r_0 + r_0 \,. \tag{7.43}$$

The latter series converges absolutely in $\mathcal{X}(A,\Sigma)$ (note that r_ν does not necessarily belong to M_2). It follows from lemma 7.3.3 that

$$\|P_0^n Q\|_{M_1} \leq c\phi(n)\|P_\nu - P_0\| \,. \tag{7.44}$$

Therefore, the series (7.43) is dominated by the numerical series

$$\sum_{\substack{k\geq 1\\ i_1,\dots,i_k\geq 0}} \phi(i_1)\cdots\phi(i_k)c^k\|P_\nu - P_0\|^k + 1 = \sum_{k=0}^{\infty}(\|P_\nu - P_0\|c\phi)^k \,. \tag{7.45}$$

The series (7.45) is convergent, provided that $\|P_\nu - P_0\| < 1/c\phi$ and, consequently, the series (7.43) is absolutely convergent. We will now prove equality (7.43). For this, we estimate the difference

$$\|\sum_\delta Q^\delta P_0^\infty - (P_0+Q)^n\|_{M_1} \,.$$

The following equality is obvious:

$$(P_0+Q)^n = P_0^n + P_0^{n-1}Q + P_0^{n-2}QP_0 + \cdots + P_0^{n-k}QP_0^{k+1} + \cdots + Q^n \,. \tag{7.46}$$

Hence

$$\begin{aligned}
\| & \sum_\delta Q^\delta P_0^\infty - (P_0+Q)^n\|_{M_1} \\
\leq & \sum_{\substack{k\geq 1\\ i_1+\cdots+i_k<n/2}} \|Q^\delta\|\|P_0^\infty - P_0^{n-i_1-\cdots-i_k}\| \\
& + \sum_{\substack{k\geq 1\\ i_1+\cdots+i_k\geq n/2}} \|Q^\delta\|\|P_0^\infty\| + \sum_{\substack{k\geq 1\\ i_1+\cdots+i_k\geq n/2}} \|Q^\delta\| \\
\leq & \sum_{\substack{k\geq 1\\ i_1+\cdots+i_k<n/2}} \|Q^\delta\|\phi(n-i_1-i_2-\cdots-i_k) + 2 \sum_{\substack{k\geq 1\\ i_1+\cdots+i_k\geq n/2}} \|Q^\delta\| \\
\leq & \sum_{\substack{k\geq 1\\ i_1+\cdots+i_k<n/2}} \|Q^\delta\|\phi(n-i_1-\cdots-i_k) + \sum_{\substack{k<\sqrt{n/2}\\ i_1+\cdots+i_k>n/2}} \|Q^\delta\| + \sum_{\substack{k\geq\sqrt{n/2}\\ i_1+\cdots+i_k\geq n/2}} \|Q^\delta\|
\end{aligned}$$

$$
\begin{aligned}
\leq & \max_{n>m>n/2} \phi(m) \sum_{k=0}^{\infty} (\|P_\nu - P_0\| 2\phi)^k \\
& + 2 \sum_{\substack{k<\sqrt{n/2} \\ i_1+\cdots+i_k>n/2}} \phi(i_1)\cdots\phi(i_k) c^k \|P_\nu - P_0\|^k \\
& + 2 \sum_{\substack{k<\sqrt{n/2} \\ i_1+\cdots+i_k>n/2}} \phi(i_1)\cdots\phi(i_k) c^k \|P_\nu - P_0\|^k \\
\leq & \max_{n>m>n/2} \phi(m) \sum_{k=0}^{\infty} (\|P_\nu - P_0\|\, 2\phi)^k \\
& + \sum_{\substack{k<\sqrt{n/2} \\ i_1+\cdots+i_k>n/2}} \phi(i_1)\cdots\phi(i_k) c^k \|P_\nu - P_0\|^k \\
& + \sum_{k>\sqrt{n/2}} (\|P_\nu - P_0\| c\phi)^k .
\end{aligned}
$$

The first and third sums in the right-hand side of (7.47) converge to zero as $n \to \infty$ because the series (7.45) is convergent and $\phi(n)$ converges to zero as $n \to \infty$. Every term of the second sum contains a factor $\phi(m)$ with $m \geq \sqrt{n/2}$. Therefore, the second sum is dominated by

$$
2 \max_{m>\sqrt{n/2}} \phi(m) \sum_{k<\sqrt{n/2}} \|P_\nu - P_0\| (c\phi)^k ,
$$

which converges to zero as $n \to \infty$. Thus (7.43) is proved.

It remains to prove the analyticity of the vector

$$
r_\nu = \lim_{n\to\infty} P_\nu^n x, \quad \text{for } x \in M_1 .
$$

The absolute convergence of (7.43) implies the analyticity of r_ν in Q for $\|Q\|$ smaller than some q_0. Since Q is an analytic function of ν for $\nu \in D$, there exists ν_2 such that, for $|\nu| < \nu_2$, we have $\|Q_\nu\| < q_0$, which implies the analyticity of r_ν in ν for $|\nu| < \nu_2$.
The proof of theorem 7.3.2 is concluded. ■

We shall make some comments. In the case of MCs satisfying Doeblin's condition (section 1.4), one can take M equal to the set of all probability measures on S. In general M consists of sufficiently smooth functions defined on the spectrum of P_0. The fact that we need two sets of uniform convergence is clear from the fact that generally $P_\nu M$ is not contained in

M for $\nu \neq 0$. But if $P_0 M_1 \subset M_1$ then M_2 can be trivially constructed, as shown below.

Remark All results of this chapter could be generalized for a general state space S, with absolutely continuous transition probabilities *à la* Tweedie [Twe76]. In this case one must take $\mathcal{X}$ equal to a suitable Banach space of measures on S.

7.4 Proof of analyticity completed

Now we finish the proof of theorem 7.1.2. For some $\alpha > 1$, we introduce the set

$$M_\alpha = \{(x_0, x_1, \ldots, x_n \ldots) : \ |x_i| \le \alpha \pi_i(0); \ \sum_{i=1}^{\infty} x_i = 1\}\,. \qquad (7.47)$$

We prove that M_α is a set of uniform convergence for the operator $P(0)$ corresponding to the Markov chain L^0. We show that $P(0)M_\alpha \subset M_\alpha$. Let $x = (x_0, x_1, \ldots, x_n) \in M_\alpha$ and set

$$P(0)x = y = (y_0, y_1, \ldots, y_n, \ldots)\,.$$

Thus

$$|y_j| \ = \ \left|\sum_{i=0}^{\infty} p_{ij}(0) x_i\right| \ \le \ \sum_{i=0}^{\infty} p_{ij}(0) \alpha \pi_i(0) = \alpha \pi_j(0)\,.$$

Moreover, $\sum_{j=0}^{\infty} y_j = 1, y \in M_\alpha$. Set

$$\phi_n = \sup_{x \in M} \|P^n(0)x - \pi(0)\|\,. \qquad (7.48)$$

We have

$$\begin{aligned}
\phi_n &= \sup_{x \in M} \sum_{j=0}^{\infty} |\sum_{i=0}^{\infty} x_i\, p_{ij}(n,0) - \pi_j(0)| \\
&= \sup_{x \in M} \sum_{j=0}^{\infty} |\sum_{i=0}^{\infty} x_i (p_{ij}(n,0) - \pi_j(0))| \\
&\le \sup_{x \in M} \sum_{i=0}^{\infty} |x_i| \sum_{j=0}^{\infty} |p_{ij}(n,0) - \pi_j(0)| \\
&\le \sum_{i=0}^{\infty} \alpha \pi_i(0) \sum_{j=0}^{\infty} |p_{ij}(n,0) - \pi_j(0)|, \qquad (7.49)
\end{aligned}$$

$$\begin{aligned}\sum_{n=1}^{\infty}\phi_n &< \alpha\sum_{n=1}^{\infty}\sum_{i=0}^{\infty}\pi_i(0)\sum_{j=0}^{\infty}|p_{ij}(n,0)-\pi_j(0)| \\ &= \alpha\sum_{i=0}^{\infty}\pi_i(0)\sum_{n=1}^{\infty}\sum_{j=0}^{\infty}|p_{ij}(n,0)-\pi_j(0)| \,. \qquad (7.50)\end{aligned}$$

To estimate the series (7.50), we use lemma 7.2.4 and inequality (7.7) of theorem 7.1.2. This yields

$$\sum_{n=1}^{\infty}\phi_n < \alpha\sum_{i=0}^{\infty}c_2\exp(-\delta_2 f_i^{\nu})[\sigma_3 f_i^{\nu} + c_7\exp(-\delta_7 f_i^{\nu})] \,. \qquad (7.51)$$

It follows from (7.51) that there exist constants a_1 and b_1 such that

$$\sum_{n=1}^{\infty}\phi_n < a_1\sum_{i=0}^{\infty}\exp(-b_1 f_i^{\nu}) \,. \qquad (7.52)$$

The convergence of the series (7.52) follows from condition (ii) of definition 7.1.1. By the same token, we have shown that M_α is a set of uniform convergence for the operator $P(0)$. Take $\alpha_1 < \alpha_2$ and the sets M_{α_1} and M_{α_2} of uniform convergence for the operator $P(0)$, such that $M_{\alpha_1} \subset M_{\alpha_2}$. Take $x = (x_1, x_2, \ldots) \in M_{\alpha_1}$. We have $\sum_1^{\infty} x_i = 1$ and $x_i \leq \alpha_1\pi_i(0)$.
Set $y = (y_1, y_2, \ldots) = P(\nu)x$, where $y_i = \sum_{j=0}^{\infty} p_{ij}(\nu)x_j$. We have $\sum_0^{\infty} y_i = 1$, since $P(\nu)$ is a Markov operator. The probability $p_{ij}(\nu)$ is different from zero only for $j \in V_i$ (the set V_i is defined in the hypothesis of the theorem). It follows from (iv) of definition 7.1.1 that there is a constant $a > 0$ such that for all points $j \in V_i$ we have

$$\sum_{j\in V_i}\pi_j(0) < a\pi_i(0) \,,$$

for any $i \in A$. Therefore

$$|y_i| \leq \sum_{j=0}^{\infty}p_{ji}(\nu)|x_j| \leq \sum_{j=0}^{\infty}p_{ji}(\nu)\alpha_1\pi_j(0) < a\alpha_1\pi_i(0) \,. \qquad (7.53)$$

It follows from (7.53) that, if $a\alpha_1 < \alpha_2$, then $y \in M_2$, and condition (ii) of theorem 7.3.2 is satisfied by the same token. Let $x^1, x^2 \in M_{\alpha_1}$. We show then there exists ν_0, such that

$$z = x^1 + \frac{(P_\nu - P_0)x^2}{\|P_\nu - P_0\|} \in M_{\alpha_2} \,,$$

for $|\nu| < \nu_0$, where $z = (z_0, z_1, \ldots)$. From the definition of z, we have $\sum_0^\infty z_i = 1$. We also show that $|z_i| < \alpha_2 \pi_i(0)$. We have

$$|z_i| \leq |x_i^1| + \frac{1}{\|P_\nu - P_0\|} \left| \sum_{j=0}^{\infty} (p_{ji}(\nu) - p_{ji}(0)) x_j^2 \right| . \tag{7.54}$$

The quantities $p_{ji}(\nu)$ and $p_{ji}(0)$ are different from zero only for $j \in V_i$. As shown before, there exists an $a > 0$ such that, for all points $j \in V_i$ and any $i \in A$, we have $\sum_{j \in V_i} \pi_j(0) < a\pi_i(0)$, so that

$$\begin{aligned} |z_i| &< \alpha_1 \pi_i(0) + \frac{1}{\|P_\nu - P_0\|} \max_{j \in V_i} |p_{ji}(\nu) - p_{ji}(0)| \sum_{j \in V_i} \alpha_1 \pi_j(0) \\ &\leq \alpha_1 \pi_i(0) + \alpha_1 a \pi_i(0) \\ &= \pi_i(0) \alpha_1 (1 + a) . \end{aligned}$$

Setting $\alpha_2 > \alpha_1(1 + a)$, we obtain that $z \in M_{\alpha_2}$. By the same token, the hypotheses of theorem 7.3.2 are satisfied. This in turn implies the analyticity of the stationary probabilities and concludes the proof of theorem 7.1.2. ■

7.5 Examples of analyticity

Consider a family $\{L^\nu\}$ of homogeneous irreducible aperiodic Markov chains with discrete time and set of states $\mathbf{Z}_+^N = \{(z_1, \ldots, z_N) : z_i > 0, \text{ integer}\}$ ($\nu \in D$, which is an open subset of the real line). We shall assume that the homogeneity condition and the condition of boundedness of jumps introduced in section 6.3 are satisfied. Besides, we assume that there exist $n, \delta > 0$ such that, for any $\nu \in D$, $\alpha \in \mathbf{Z}_+^N$ and $\beta \in V_\alpha$, where

$$V_\alpha = \{\beta : \sup_{\nu \in D} p_{\alpha\beta}(\nu) > 0\},$$

we have

$$p_{\beta\alpha}(n, \nu) > \delta .$$

As in section 6.3, we introduce the family of vector fields $\{V^\nu, \nu \geq 0\}$.

Theorem 7.5.1 *Assume that the operator P_ν defined by the chain L^ν depends on ν analytically for all $\nu \in D$. If there exists a set $U \subset D$ such that the family of vector fields $\{V^\nu, \nu \in D\}$ satisfies condition $\tilde{B}$ of section 6.3, then the chains L^ν are ergodic, for all $\nu \in U$, and the* *stationary probabilities $\pi_\alpha(\nu)$ are analytic in ν, for any $\alpha \in \mathbf{Z}_+^N$ and $\nu \in U$.*

Proof As in section 6.3, condition $\tilde{B}$ implies the existence of a function $f^\nu(\alpha), \nu \in D$, which satisfies (7.6). Moreover, the families of functions $\{k_i^\nu\}$ and $\{f_i^\nu\}$ satisfy all necessary conditions to be analytic Lyapounov families (see definition 7.1.1). Therefore, it follows from theorem 7.1.2 that the $\pi_\alpha(\nu)$'s are analytic with respect to ν, for $\alpha \in \mathbf{Z}_+^N$ and all $\nu \in U \subset D$.
For the families of random walks $\{L^\nu\}$ in $\mathbf{Z}_+^N$, with $N \leq 3$, the analyticity conditions for the stationary probabilities can be formulated explicitly, since we have succeeded in constructing a function $f(\alpha)$ satisfying condition $\tilde{B}$ for the family $\{L^\nu\}$. Let us formulate the precise theorem for random walks in $\mathbf{Z}_+^3$.

Theorem 7.5.2 *Assume that the Markov chain L^{ν_0}, where $\nu_0 \in D$, satisfies the hypotheses of theorem 4.4.4 guaranteeing the ergodicity of L^ν. Moreover, assume that the operator P_ν depends analytically on ν for $\nu \in D$. Then there is a neighbourhood U of ν_0, $U \subset D$, such that the chains L^ν are ergodic, for all $\nu \in U$, and the stationary probabilities $\pi_\alpha(\nu)$ are analytic in ν, for any $\alpha \in \mathbf{Z}_+^3$ and all $\nu \in U$.*

The proof of this theorem mimics totally that of theorem 6.3.2 and will not be included.
Other examples of analytic Lyapounov families are Jackson networks, which were introduced in section 3.5. Then, using the notation of section 3.5, theorems 3.5.7 and 3.5.8 can be rewritten in terms of analytic Lyapounov families as follows:

Theorem 7.5.3 *For the Lyapounov function f_x satisfying the conditions of theorem 3.5.7 with $k(x) \equiv 1$, the Jackson network is an analytic Lyapounov Markov chain.*

Theorem 7.5.4 *Let us consider a Jackson network such that $0 \in \Pi$. Then, for the Lyapounov function f_x satisfying the conditions of theorem 3.5.8 with $k(x) \equiv k$ sufficiently large, the Jackson network is an analytic Lyapounov Markov chain.*

Bibliography

[Afa87] L.G. Afanas'eva. On the ergodicity of an open queueing network. *Theory of probability and applications*, 32(4):777–781, 1987.

[AFM] I. Asymont, G. Fayolle, and M. Menshikov. Random walks in a quarter plane with zero drifts. II: Transience and Recurrence. To appear.

[AIM] I. Asymont, R. Iasnogorodski, and M. Menshikov. Random walks with asymptotically zero drifts. To appear.

[BFK92] A.A. Borovkov, G. Fayolle, and D.A. Korshunov. Transient phenomena for Markov chains and applications. *Adv. Appl. Prob.*, 24(3), 1992. (Rapport de Recherche INRIA 1171. Mars 1990).

[BM89] F. Baccelli and A.M. Makovski. Queueing models for systems with synchronization constraints. *Proceedings of the IEEE*, 77(1):138–161, 1989.

[Bor86] A.A. Borovkov. Limit theorems for queueing networks I. *Theory of probability and applications*, 31(31):474–490, 1986.

[Chu67] K.L. Chung. *Markov chains with stationary transition probabilities.* Springer, 1967.

[Fay89] G. Fayolle. On random walk arising in queueing systems : ergodicity and transience via quadratic forms as lyapounov functions - part I. *Queueing Systems*, 5:167–184, 1989.

[FB88] G. Fayolle and M.A. Brun. On a system with impatience and repeated calls. *CWI monographs. Liber Amicorum for J.W. Cohen*, 7:283–305, 1988.

[Fel56] W. Feller. *Boundaries induced by positive matrices.* Volume 83, Trans. Amer. Math. Soc., 1956. pp. 19-54.

[FI79] G. Fayolle and R. Iasnogorodski. Two coupled processors : the reduction to a Riemann-Hilbert problem. *Z. Wahrsch. verw. Gebiete*, 47:325–351, 1979.

[Fil89] Yu.P. Filonov. Ergodicity criteria for homogeneous discrete Markov chains. *Ukrainian Math. J.*, 41(10):1421–1422, 1989.

[FIVM91] G. Fayolle, I Ignatyuk, V.A. V.A.Malyshev, and M.V. Menshikov. Random walks in two-dimensional complexes. *Queueing Systems*, 9(3):269–300, 1991.

[FMM92] G. Fayolle, V.A. Malyshev, and M.V. Menshikov. Randoms walks in a quarter plane with zero drifts. I. Ergodicity and Null Recurrence. *Ann. Institut Henri Poincaré, Probabilités et Statistiques*, 28(2):179–194, 1992.

Rapport de Recherche INRIA N^0 1314, 1990.

[Fos53] F.G. Foster. On stochastic matrices associated with certain dueueing processes. *Ann. Math. Stat.*, 24:355–360, 1953.

[Fos89] S.G. Foss. Some properties of open queueing systems. *Problems of information transmission*, 25(3), 1989.

[GS74] I.I Gihman and A.V Skorohod. *The Theory of Stochastic Processes.* Springer-Verlag Berlin Heidelberg New-York, 1974. Trnslated from Russian. (Nauka, Moscow, 1971).

[ICB88] G Iazeolla, P.J Courtois, and O.J Boxma, editors. *Transaction processing time in a simple fork-join system*, North-Holland, Elsevier Science Publishers, 1988. Proc. of the Second Int. MCPR Workshop, Rome, Italy, May 25-29, 1987.

[Jac63] J.R. Jackson. Jobshop-like queueing systems. *Management Science*, 10:131–142, 1963.

[Kal73] V.V. Kalashnikov. The property of γ-reflexivity for Markov sequences. *Dokl. Akad. Nauk. SSSR*, 213:1243–1246, 1973.

[Kar68] S. Karlin. *A first course in stochastic processes.* Academic Press, 1968.

[Key84] E.S. Key. Recurrence and transience criteria for random walks in a random environement. *The Annals of Probability*, 12(2):529–560, 1984.

[KKR88] M.Ya. Kelbert, M.L. Kontsevich, and A.N. Rybko. Jackson networks on countable graphs. *Theory of probability and applications*, 33(2):358–361, 1988.

[Kor90] D.A. Korshunov. *Transient phenomena for real-valued Markov chains.* PhD thesis, Institute of Mathematics, Novosibirsk, 1990.

[KSF80] I.P. Kornfeld, Ya.G. Sinai, and S.V. Fomin. *Ergodic Theory.* Springer Verlag, Moscow, 1980.

[Lam60] J. Lamperti. Criteria for the recurrence or transience of stochastic process I. *Journal of Math. Anal. and Appl.*, 1:314–330, 1960.

[Lam63] J. Lamperti. Criteria for stochastic processes II. Passage time moments. *Journal of Math. Anal. and Appl.*, 7:127–145, 1963.

[Mal70] V.A. Malyshev. Random walks. The Wiener-Hopf equations in quadrant of the plane. Galois automorphisms. *Moscow State University Press.*, 1970.

[Mal72a] V.A. Malyshev. Classification of two-dimensional random walks and almost linear semi-martingales. *Dokl. Akad. Nauk. USSR*, 202(3):526–528, 1972.

[Mal72b] V.A Malyshev. Homogeneous random walks in a half-strip. *Veroyatnostye metody issledovania. A.N Kolmogorov (ed.). Moscow Univ. Press*, 5–13, 1972.

[Mal73] V.A. Malyshev. Doctoral thesis. *Moscow University*, 1973.

[Mal93] V.A. Malyshev. Networks and dynamical systems. *Adv. in Appl. Prob.*, 25:140–175, 1993.

[MM79] V.A. Malyshev and M.V. Menshikov. Ergodicity, continuity and analyticity of countable Markov chains. *Trans. Moscow. Math Soc.*, 39:3–48, 1979.

[MSZ78] J.F. Mertens, E. Samuel-Cahn, and S. Zamir. Necessary and sufficient conditions for recurrence and transience of Markov chains in terms of inequalities. *J. Applied Probability*, 15:848–851, 1978.

[Nau89] H. Nauta. *Ergodicity conditions for a class of two-dimensional queueing problems.* PhD thesis, Utrecht University, January 1989.

[Nev72] J. Neveu. *Martingales a temps discret.* Masson, 1972.

[Pod85] V. Podorolski. *Diploma thesis in probability.* Master's thesis, Moscow State University, 1985.

[Pop77] N.N Popov. Conditions for geometric ergodicity of countable Markov chains. *Soviet Math. Dokl.*, 18:676–679, 1977.

[Pro56] Yu.V. Prohorov. Convergence of random processes and limit theorems of probability theory. *Teor. Veroyatnost. i. Primenen 1*, 117–238, 1956.

[RW88] M.I. Reiman and R.J. Williams. A boundary property of semimartingale reflecting brownian motions. *Prob. Theor. Rel. Fields*, 77(87–97), 1988.

[Szp90] W. Szpankowski. Towards computable stability criteria for some multidimensional stochastic processes. *In Stochastic Analysis of Computer and Communication Systems*, 1990. H. Takagi (Editor), North Holland.

[Twe76] R.L. Tweedie. Criteria for classifying general Markov chains. *Adv. Appl. Prob.*, 8:737–771, 1976.

[VW85] S.R.S. Varadhan and R.J. Williams. Brownian motion in a wedge with oblique reflection. *Commun. on Pure Applied Mathematics*, 38:405–443, 1985.

[Wil85] R.J. Williams. Recurrence classification and invariant measure for reflected brownian motion in a wedge. *Ann. Prob.*, 13(3):758–778, 1985.

Index

almost linear, 62
analytic approach, 57
analytic family of Markov chains, 149
analytic Lyapounov families, 164
analytic Lyapounov family, 149
aperiodic, 7
approach via diffusion processes, 57

billiard dynamical systems, 86
Birkhoff's ergodic theorem, 86
Birth and death process, 12
Boundedness condition, 141
Boundedness of moments, 34
Boundedness of the jumps, 37, 100, 145
bristle, 103
Buffered ALOHA, 38

communicate, 7
Condition χ, 133
Condition $\bar{B}$, 145
Condition λ, 132
coupled-processors, 38

deterministic, 92
discrete time homogeneous Markov chain (MC), 6
Doeblin's condition, 11, 150

embedded, 10
ergodic, 8, 29, 30, 34, 40, 41, 58, 60, 61, 64, 65, 73, 74, 80, 81, 90, 93, 94, 101, 107, 111
ergodic face, 84
ergodic faces, 82
essential, 7
essential class, 7
exit boundary problem, 104

face, 37, 64
first generation planes, 85
First moment condition, 39
first moment condition, 92
first vector field, 38, 65
Fork-Join, 38
Foster's criterion, 35, 60
Foster's Theorem, 29

γ-recurrence, 141
geometrically ergodic, 156

hedgehog, 101
Homogeneity, 33
Homogeneity condition, 145

induced chain, 34
induced Markov chain, 79
inessential state, 7
ingoing bristle, 103
ingoing face, 80, 101
intensity matrix, 10
interaction, 38
irreducible, 7

Jackson network, 65
Jackson networks, 38, 62, 164
Jackson's system, 63
Jackson's theorem, 64
jump times, 10

Lower boundedness, 33, 39
Lower boundedness of the jump, 92
Lyapounov function, 16
Lyapounov functions, 33

martingale, 16
Maximal homogeneity, 37
Maximal space homogeneity, 100
maximal space homogeneity, 39
method of Lyapounov functions, 57

neutral face, 80
non ergodic, 8
non recurrent, 8
non-ergodic, 30, 40, 58, 73, 74, 76, 80
non-zero assumptions, 100
null recurrence, 61
null recurrent, 8, 31, 41, 58

orthogonal projection, 102
outgoing face, 80, 101

Partial homogeneity, 92
path, 6
path space, 6
periodic random walk, 36
piecewise linear, 62
planar, 99
point regular, 83
point stable, 84
Popov's Theorem, 156
position of the chain, 6
positive recurrent, 8, 35
principle of almost linearity, 69
principle of local linearity, 45, 46, 55

Quasi-deterministic process, 108
Queueing, 123

random element φ, 5
random variable, 5
random walk deterministic, 86
random walk essentially deterministic, 86
random walk regular, 83
random walk strongly regular, 84
recurrent, 8, 26, 34, 36, 61, 73, 76, 106

scattering probability, 103
second generation planes, 85
Second moment condition, 56
second vector field, 81, 102
semi-analytic approach, 57
set of uniform convergence, 157, 161
smallness, 62
smoothing, 69
stability, 76
strongly connected, 100
submartingale, 58
surface, 67

third generation planes, 85
transient, 8, 28, 31, 36, 41, 61, 73, 75, 76, 80, 90, 93
Tweedie, 161
two-dimensional complex, 99

value of the chain, 6

x-bundle, 83

Zero drifts, 56

Printed in the United States
By Bookmasters